Animals & Men

A COLONY OF FIRE SALAMANDERS IN NORTHUMBERLAND?
An interview with Wes Sullivan; Sea Serpents;
Hunting the Blue Devil; Butterfly invaders; the
science of Owlman and more

The Journal of the Centre for Fortean Zoology; Issue 53 - May 2015

Contents

Typeset by Jonathan Downes,
Cover and Layout by SPiderKaT for CFZ Communications
Using Microsoft Word 2000, Microsoft Publisher 2000, Adobe Photoshop CS.
First published in Great Britain by CFZ Press

CFZ Press, Myrtle Cottage, Woolsery, Bideford, North Devon, EX39 5QR

© CFZ MMXV

ISBN: 978-1-909488-31-1

Faculty of the Centre for Fortean Zoology

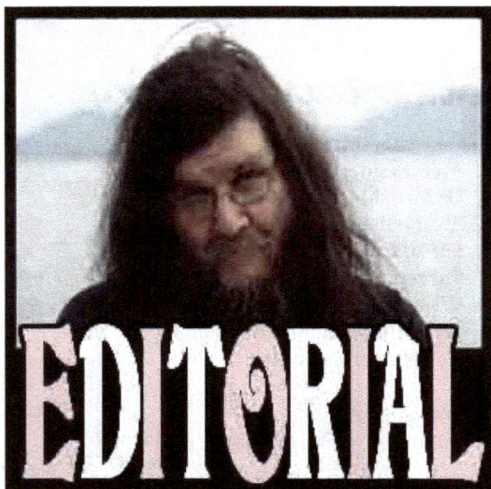

EDITORIAL

Dear Friends,

Welcome to the second issue of *Animals & Men* for 2015. At the beginning of the year I promised that this would be the year during which we got back to our intended four issue publishing schedule, for the first time (and I blush to add it this) since 2003.

I also intend to get back to publishing the CFZ Yearbook on an annual basis, and - as promised in the last issue - to presenting CFZ members with twelve monthly electronic newsletters. Things have not gone quite the way I intended this year; some of the other refurbishment of the CFZ has taken longer and cost more than I had originally envisaged. The monthly newsletters will happen, I promise, and I would like to think that they will arrive before the next issue of this magazine.

There will probably not be a yearbook this year, but I live in hope that we will get one for 2016. The latest volume of *The Journal of Cryptozoology* (Volume Three) has just been released, and personally I think it is the best yet. Something else that has fallen into abeyance in recent months is our monthly webTV show *On The Track* but I live in hope that it will return in due course.

We have ordered the building materials needed to repair the museum roof, and Graham will - I sincerely hope - be replacing the current damaged roof sometime next week, whereupon both the conservatory and the museum can return to the sort of state that they were a couple of years ago. The current CFZ workforce includes two hardworking teenage girls and an equally hardworking teenage boy, and I am sure that once I set them loose with whitewash, paintbrush and hoover, we shall have the museum looking shipshape, and rather snazzy again well in time for the Weird Weekend.

Talking about the Weird Weekend, this year's event is once again scheduled for the Small School in Hartland, on the third weekend of August. It will be our sixteenth annual event, and the second at the school. All food and drink sold during the weekend will raise money for this singular and very valuable educational establishment. I can't actually enforce this, but I would ask people not to bring their own food and drink with them, but to support the school by buying theirs.

Andrew May has done a remarkable job dragging us kicking and screaming into the 21st Century, and we not only have a swish new website for The CFZ Publication Group, but also an impressive range of ebooks, which is expanding all the time. Andrew has also taken an interest in our Fortean Fiction imprint and is working hard to expand it. Ronan Coghlan has also done sterling work on the CFZ Publishing Facebook page, and has, today, agreed to take over more publicity duties. I have never pretended to be a businessman, and the CFZ is only a business in the vaguest, and most surreal and esoteric terms, but we do live in a capitalist society (much though I deplore both the vulgarity and the ethics of it) so I thank everyone who is pulling together to help us keep our financial heads above water.

The project that I most want to achieve is a standalone CFZ Library. We do, of course, have a library, but at the moment it is doubling as Mother's bedroom, and it will continue to do so for the foreseeable future, and quite probably fir the rest of the old lady's life. Back in 2010, Richie and Naomi West took us to the Anomaly Archives in Austin,

The Great Days of Zoology are not done!

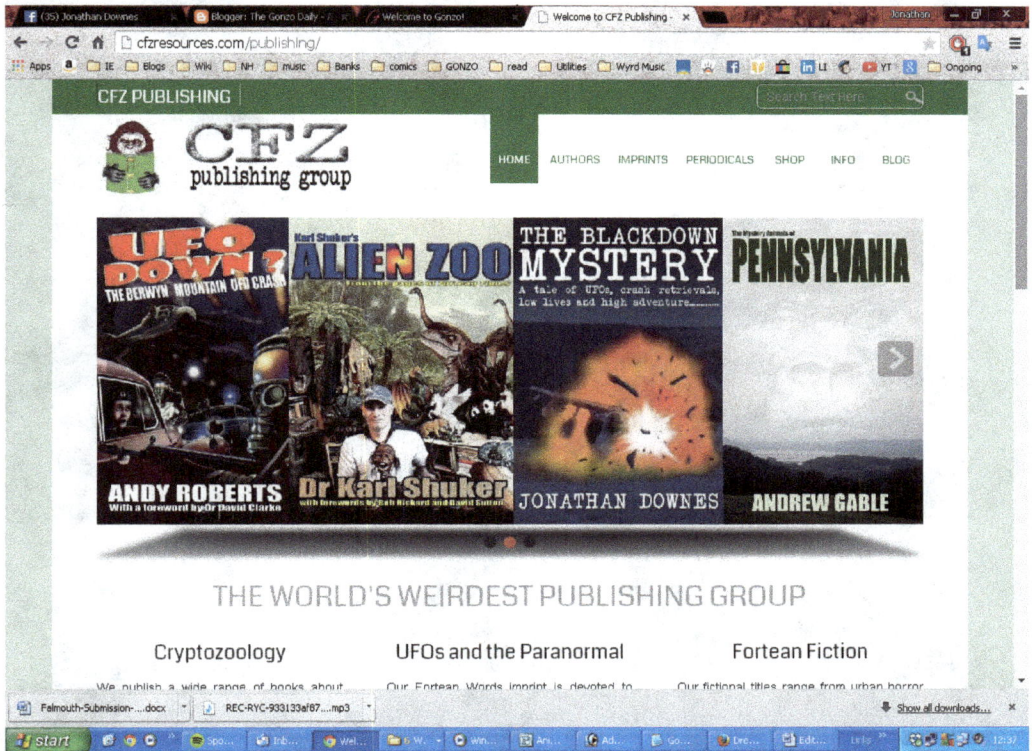

TX, run by S Miles Lewis. I was highly impressed. The organisers have taken what is basically a unit in an industrial estate and converted it into a highly impressive library and research centre focussing on esoteric subjects. I took one look at it and decided that I wanted to do something similar. I am hoping that we can get hold of an industrial unit somewhere in North Devon, and have it staffed by pupils from the school, so that after all operating costs have been covered, all profits can be ploughed back into the school itself.

I have been asked on a number of occasions why I have become so enthusiastic a supporter of the Small School so quickly. The answer is simple. My schooldays were very unhappy ones, and I garnered next to nothing from them except a deep distrust and suspicion of authority figures, and the knowledge that I was more intelligent than most of the idiots who were trying to teach me to conform to their

outdated shibboleths. But at least they were trying to teach me *something*. These days the teaching profession in the United Kingdom seems to be almost entirely about ticking the appropriate boxes on OFSTED report forms, and teaching pupils to pass exams rather than actually instilling into them any love of learning for learning's sake. The Small School is refreshingly different in this respect, and so I shall continue to do what I can to support them.

The CFZ has come a long way in the twenty three years we have been in operation, and we have some big plans for the future. Thank you very much my friends for your continuing support. It means a lot to me, as we push forward into the uncharted waters of the 21st Century.

Yours, as always

Jon Downes

THE CENTRE FOR FORTEAN ZOOLOGY
www.cfz.org.uk

A LEGAL MATTER

Newsfile

New & Rediscovered

Boxing the Right Ticks

A new tick species found in Malaysia and Vietnam was recently discovered by researchers in Georgia. Adults of the new species, *Dermacentor limbooliati*, are similar to those of *Dermacentor auratus* and of *Dermacentor compactus*, species with which it was previously confused. However, *D. limbooliati* can be distinguished by a number of characteristics which are described in an article in the *Journal of Medical Entomology* called "Description of New *Dermacentor* (Acari: *Ixodidae*) Species From Malaysia and Vietnam."

The authors, Drs. Dmitry and Maria Apanaskevich of Georgia Southern University, discovered the new species while re-examining the extensive holdings of Oriental *Dermacentor* ticks in the United States National Tick Collection (USNTC). Me? I am just amused to find that there *is* a National Tick Collection complete with its own acronym.

Source: http://www.sciencedaily.com/ releases/2015/03/150310105315.htm

Meet *Cyranorogas depardieui*. It's a wasp, but not like any wasp you've ever seen. Its face has a sharp ridge, like the blade of a knife, sticking out of it. The new wasp was discovered during an expedition to Mount Wilhelm, the highest mountain in Papua New Guinea, in 2012.

It has now been formally described in the *Journal of Natural History* by Buntika Butcher and Donald Quicke of Chulalongkorn University in Bangkok, Thailand. "The wasp is very unusual," says Quicke. The species name contains two jokes about big noses. "Cyranorogas" refers to Cyrano de Bergerac, a 17th-century French playwright who supposedly had a large nose, while "depardieui" refers to the actor Gérard Depardieu, who played Bergerac in a 1990 film.

SOURCE: http://www.bbc.com/earth/story/20150306-the-wasp-with-a-face-like-a-knife

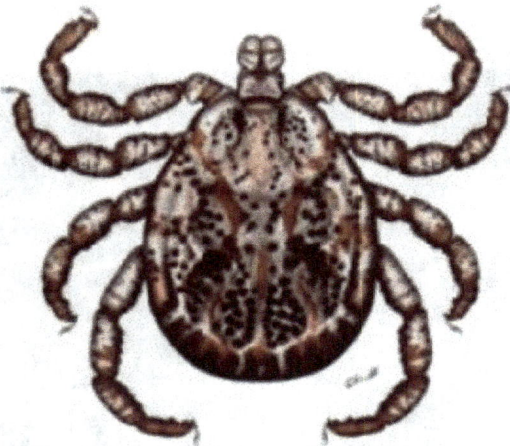

© Entomological Society of America

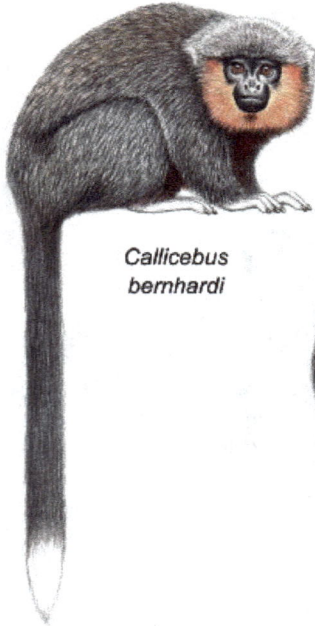

Callicebus
bernhardi

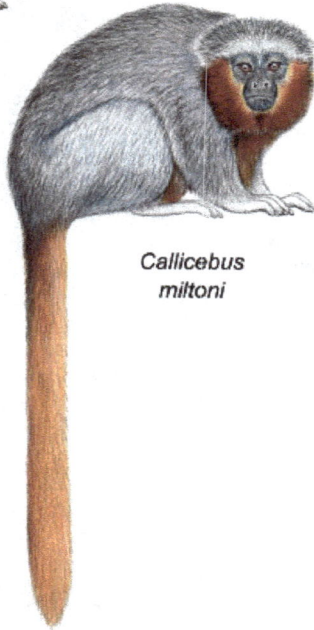

Callicebus
miltoni

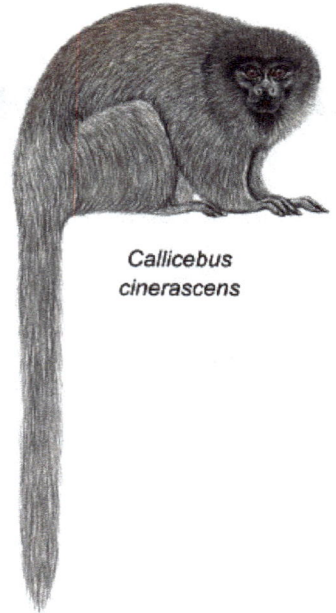

Callicebus
cinerascens

Too much monkey business

A new species of titi monkey has been discovered by scientists in Brazil. Titis are new world monkeys found across South America. These tree-dwelling primates have long, soft fur and live in small family groups consisting of a monogamous pair and their offspring. Rather touchingly, they are often observed sitting or sleeping with their tails entwined. After researcher Julio César Dalponte spotted an unusual looking titi monkey on the east bank of

the Roosevelt River, whose colouration did not match any known species in 2011 a team of scientists supported by the Conservation Leadership Programme decided to investigate.

Over the course of a number of expeditions, the team recorded several groups of these unusual monkeys, whose ochre sideburns, bright orange tail and light grey forehead stripe set them apart from other known species in the genus. They have named *Callicebus miltoni* (or Milton's titi monkey) in honour of Dr Milton Thiago de Mello, a noted Brazilian primatologist who is credited with training many of the country's top primate experts. Because they are not able to swim or cross mountainous terrain, these monkeys are restricted to a small area, effectively hemmed in by a number of rivers and hills. This small range could put the species at risk from human activities, particularly because only around a quarter of this area is protected.

SOURCE: http://www.wildlifeextra.com/go/news/ new-monkey-species.html#cr

Here be Dragons

While researching the two known species of seadragons as part of an effort to understand and protect the exotic and delicate fish, scientists at Scripps Institution of Oceanography at UC San Diego made a startling discovery: A third species of seadragon.

Using DNA and anatomical research tools, Scripps graduate student Josefin Stiller and marine biologists Nerida Wilson of the Western Australia Museum (WAM) and Greg Rouse of Scripps Oceanography found evidence for the new species while analyzing tissue samples supplied by WAM. The researchers then requested the full specimen as well as photographs taken just after it was retrieved from the wild in 2007. They were further surprised by the appearance of the newly identified animal. The color was a bright shade of red and vastly different from the orange tint in Leafy Seadragons and the yellow and purple hues of Common Seadragons.

Stiller, Wilson, and Rouse gave their new discovery the scientific name *Phyllopteryx dewysea*, also referred to as the "Ruby Seadragon," and details are published in the journal Royal Society Open Science.

SOURCE: https://scripps.ucsd.edu/news/new- species-ruby-seadragon-discovered-scripps- researchers

Weird wallaby

Rock-wallabies (genus Petrogale) are small to medium-sized marsupials, weighing from 1 to 12 kg.

These animals are only found in mainland Australia and some offshore islands, being absent from Tasmania and New Guinea.

They represent one of the largest groups of extant macropods (kangaroos, wallabies and their relatives) distributed across the country,

where they inhabit complex rocky environments such as cliffs, gorges, outcrops and escarpments.

In a new DNA study, Dr Potter and her colleagues found that two populations of a widespread and common species, the short-eared rock-wallaby (*Petrogale brachyotis*), – one from the Kimberley and western Northern Territory, the other from the northern and eastern Northern Territory – are genetically distinct species. The scientists said that members of the latter population are not only smaller (2.6 – 3.5 kg), but differ in coloration and markings being predominately dark grey/brown, with distinct head and side stripes, as well as brightly coloured limbs. In contrast, the short-eared rock-wallaby from the Kimberley is larger (3.9 – 4.5 kg), lighter and greyer, with much less prominent marking. This discovery means there are now 17 known species of rock-wallabies.

The new species is named the Wilkins' rock-wallaby (*Petrogale wilkinsi*), after Australian explorer, naturalist and aviator Capt Sir George Hubert Wilkins who, in 1925, collected one of the first specimens in southwestern Arnhem Land. Another common name suggested for Petrogale wilkinsi is the eastern short-eared rock-wallaby.

SOURCE: http://www.sci-news.com/biology/science-petrogale-wilkinsi-wilkins-rock-wallaby-new-species-australia-02494.html

NEW FROG: *Bangladesh*

A new species of frog has hopped onto the radar of researchers in Bangladesh. The frogs were discovered after the researchers noticed their unusual breeding habits, according to a new study.

Most frogs have a specific mating season, but researchers found that one frog bred all year long, even in the winter, said study lead researcher M. Sajid Ali Howlader, a doctoral student of biosciences at the University of Helsinki in Finland.

Howlader learned that the frog was named *Euphlyctis cyanophlycti*, and it was discovered by the German naturalist J. G. Schneider in 1799. But, a detailed study of the frog's genetics, shape and size showed that it was actually an entirely different species from *E. cyanophlycti*.

He named the new 1.5-inch-long (3.8 centimetres) frog *Euphlyctis kalasgramensis*, after the Bangladesh village of Kalasgram, where he first found the frogs.

SOURCE: http://www.livescience.com/49698-new-frog-species-bangladesh.html

NEW FROG: *Costa Rica*

Already dubbed a real-life Kermit, a new species of frog has been identified in the rainforests of Costa Rica. The inch-long creature, scientifically named *Hyalinobatrachium dianae*, joins Costa Rica's 13 other glass frogs, named for their translucent bodies through which you can view their organs. (Not all glass frogs, however, sport such translucent undersides.) Despite its bright-green skin and bulging white eyes, *H. dianae* had evaded biologists until a few specimens were collected by scientists with the Costa Rican Amphibian Research Center.

SOURCE: http://www.livescience.com/50548-kermit-the-frog-look-alike-discovered.html

NEW FROG: *Madagascar*

For 20 years these tiny frogs have been sat without a scientific name. Until now. Meet *Stumpffia kibomena* – Madagascar's newest species of frog. This bright-bellied frog gets its name from the Malagasy words 'kibo' meaning 'belly' and 'mena' meaning 'red'. The new species has been described in a paper published in *Zootaxa*.

Lead author, Dr Frank Glaw from the Zoologische Staatssammlung in Munich, Germany recalled his first encounter: "When I found *Stumpffia kibomena* for the first time in early 1995, I was very excited about this beautiful discovery. "Due to its distinctive colouration with its bright red belly it was immediately clear to me, that this was a new species."

SOURCE: http://www.bbc.com/earth/story/20150330-meet-a-new-madagascar-frog-species-called-red-belly

NEW FROG: Ecuador

A frog in Ecuador's western Andean cloud forest changes skin texture in minutes, appearing to mimic the texture it sits on. Originally discovered by a Case Western Reserve University PhD student and her husband, a projects manager at Cleveland Metroparks' Natural Resources Division, the amphibian is believed to be the first known to have this shape-shifting capability. But the new species, called *Pristimantis mutabilis*, or mutable rainfrog, has company. Colleagues recently found that a known relative of the frog shares the same texture-changing quality.

SOURCE: http://www.sciencedaily.com/releases/2015/03/150323132854.htm

This image shows skin texture variation in one individual frog (*Pristimantis mutabilis*) from Reserva Las Gralarias. Note how skin texture shifts from highly tubercular to almost smooth; also note the relative size of the tubercles on the eyelid, lower lip, dorsum and limbs.
Credit: *Zoological Journal of the Linnean Society*

© cheng li, imaging biodiversity expedition / tibet forestry

NEW MACACQUE: *Want Tibet?*

Amazing images taken mainly by camera traps reveal a previously unrecognised species of macaque living in Tibet's forests. The photographs, provide an intimate portrait of the cute primates – showing the monkeys grooming each other; foraging and tending to their young. The elusive monkey, which researchers have called the white-cheeked macaque (*Macaca leucogenys*), had previously been identified as the Assamese macaque (*Macaca assamensis*). But now researchers in China who took the images say the charismatic creature is a species in its own right.

SOURCE: http://www.bbc.com/earth/story/20150408-amazing-images-reveal-new-macaque

Back from the brink

Until this year, scientists hadn't seen the Bouvier's red colobus monkey in the wild since the 1970s. It lives in groups in swampy forests along the Congo River, in the Republic of the Congo. Hunting and logging decimated its population, leading some scientists to suggest the monkey was extinct. Now, independent explorers have rediscovered the rare monkey. The researchers, Lieven Devreese of Belgium and Gaël Elie Gnondo Gobolo of the Republic of the Congo, set off in February to track down the elusive species. Their expedition was supported by donations collected through the crowdfunding website Indiegogo, and funding from the Wildlife Conservation Society.

SOURCE: http://www.livescience.com/50513-red-colobus-monkey-rediscovered-photo.html

Chupacabras

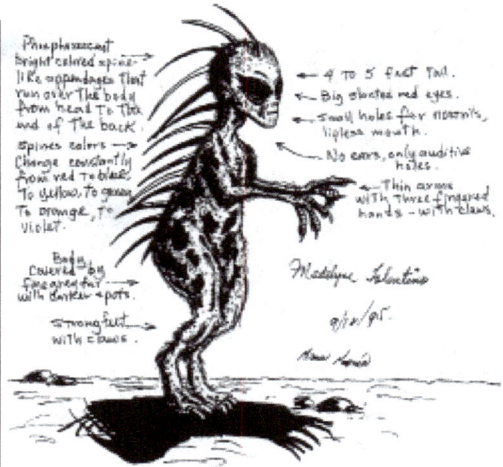

There have been a number of records over the past few months of what have been dubbed 'chupacabras'. Interestingly, the animals that we present in this section are all completely unlike each other. For the record I don't think any of these have anything to do with the original semi bipedal creatures (pictured opposite in an iconic drawing first presented by Jorge Martin in the mid 1990s, reported from the Canovenas plateau of Puerto Rico about twenty years ago.

However, as I have written in these pages, and elsewhere, it is interesting to note how the word 'chupacabras' has become a catch-all term for mysterious (and not so mysterious) animals from across Latin America. In this round up we go first to Chile.

A farmer has claimed he has found the remains of a two Chupacabra creatures also known as blood sucking monsters after stumbling across bizarre-looking dead animals in Chile. Goat farmer Javier Prohens was having lunch with friends when terrified farmhand 54-year-old Bricio Saldivar alerted them to the bodies of the strange creatures on the outskirts of the small town of Monte Patria in the east central Chilean province Limari. At first the group thought the remains were those of bats but on closer inspection, they became convinced they are from they mythical creatures.

POSTSCRIPT: Various internet sources have suggested that this noisome object is actually a mummified cat. I tend to agree but would welcome submissions and suggestions from the *Animals & Men* readership as to its identity. Sufficient to say that I have not heard any results of the putative analysis. JD

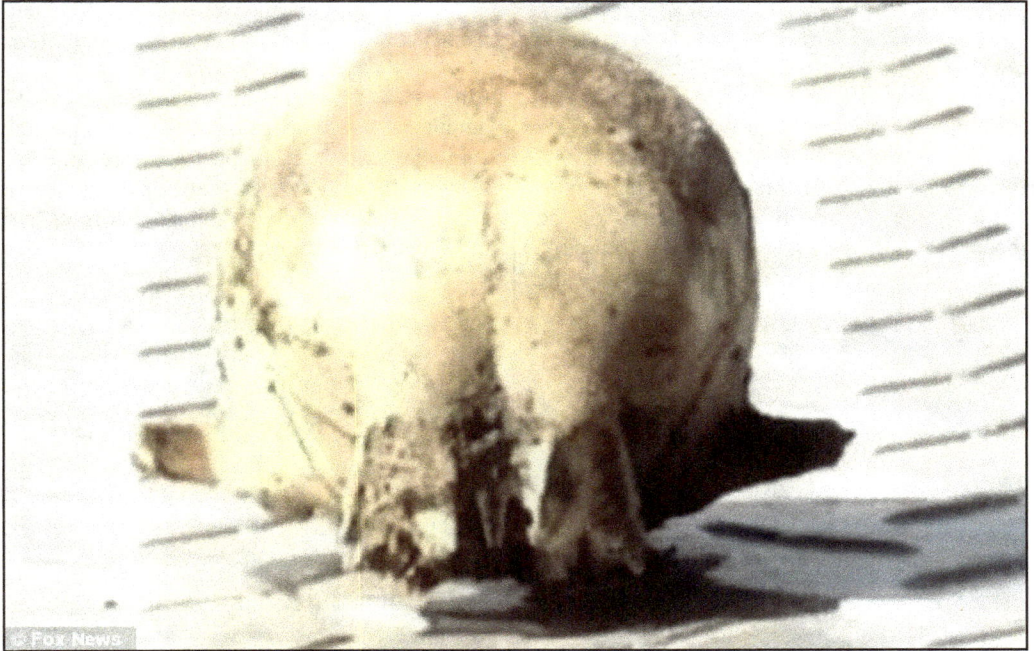

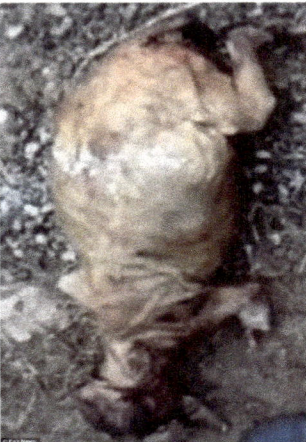

An Oklahoma woman believes she has secured photographic evidence of the legendary blood-sucking beast, el chupacabra. Vonda Thedford, 55, told Fox News that she was driving along a Pittsburg County rural road earlier this month when she spotted the mysterious carcass lying on the ground. When she stopped to look at the dead creature, she was disturbed to see it had 'a little truck' in place of a nose, 'little toes' and 'hair on its tail'. In a bid to document the unusual-looking beast, she whipped out her camera phone.

The restaurant worker says the images have left people baffled and no-one has been able to identify the bloated and hairless critter. 'I know it's something logical,' Thedford said, 'I know it's not an alien, I know it's not Bigfoot's aborted child, like people are coming up with.' Desperate to unravel the mystery, Thedford revisited the dead animal. As the flesh had rotted away, she took the remaining bones with her. They have been preliminarily identified as the bones of a young dog, but the pictures are too blurred for us to make anything approaching a definitive identification.

SOURCE: http://www.dailymail.co.uk/news/article-3054466/I-know-s-not-alien-Driver-claims-dead-chupacabra-Oklahoma-roadside-photos-prove-

Blue Dogs

Naomi West wrote: "This was taken last year around Barstow, California, from a man that used to work with Richie. Notice it appears to be a female but it also has those pads on the back." Well this is a game hanger! Until now, I have always taken it as gospel that only the male blue dogs have the weird pouches on their haunches.

But, as she says, this photograph does definitely appear to be a female; no external genitalia, and, what appear to be nipples hanging down. I would hazard a guess that this is not just a female, but a female that is, or has recently been, nursing.

Man Beasts (BHM)

Bathing Bigfoot

Possibly the clearest photograph I have ever seen purporting to be a North American mystery primate, surfaced recently. If it is real it is outstanding, if not, it is a clever photoshop job.

66 year old John Rodriguez, a retired electrician, claims to have encountered Bigfoot on December 26 on the Hillsborough River just northeast of Tampa, Florida.

He recounts: "I fish for gar in the river and I bring my camera to take pictures of the birds and what not. I heard a squishing sound, looked over and saw this thing walking through the water and crouch down in the duck weed. It did not look like a guy in a suit - it was definitely an animal. I took this picture and got out of there as fast as I could."

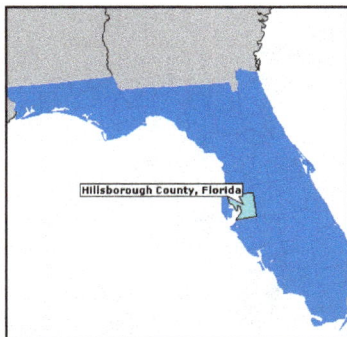

Hillsborough County, Florida

"I've heard of Skunk Ape prints around Green Swamp [in Florida], but never anything like this," Rodriguez told HuffPost in an email. "My whole life, never seen anything like it."

SOURCE: http://www.crossmap.com/news/bigfoot-sightings-2015-new-sightings-in-the-florida-green-swamp-15811

John Rodriguez

Nebraska news

Sharon Jorgensen and her border collie were walking on the east side of the golf course in Curtis, Nebraska in mid January. They

Curtis, NE

reached the cart path that climbs through the trees from the third fairway, she found the tracks that the something else had left behind. They were big.

She dropped her glove alongside one and snapped a photo; the four-toed footprint was half again as large.

They were far between. She estimated the distance between them at nearly 6 feet, an Olympic-sprint stride. They were alone. The snow surrounding them was untouched, unblemished.

"There were no other feet prints out there

but mine and this big guy," Jorgensen said. "The snow tells the story. What it is, I don't know."

"This isn't a joke," she said. "This is the middle of nowhere. If you were going to make a joke, you'd put them where someone was going to find them. This isn't that place."

Word soon got to golf course superintendent Eric Senff.

"I was actually home that morning when Sharon sent me the pictures," he said. "I went out there and looked at them." The 30-year-old has run the golf course for a half-dozen years, but until he followed the tracks from the third to the fourth, he'd never encountered anything out of the ordinary. "As big as they were and as long as the strides were, I've never seen anything like it."

He showed them to a biologist, who said they could have been made by a jackrabbit, but Senff had never seen one on the course. He remembered a farmer nearby had reported seeing a black bear a year or so ago, but that didn't fit. And he heard another farmer in the area raises emus, but he looked at pictures of their tracks online, and they weren't quite right, either.

Senff doesn't believe in bigfoot or sasquatch. He's stumped.

"I just have no idea."

SOURCE:
http://journalstar.com/news/state-and-regional/nebraska/simply-a-mystery----strange-tracks-in-snow/article_6775820b-ffba-52c6-824a-90e7a61ecb35.html

More Problems for Bryan Sykes

Bryan Sykes has been forced to issue a retraction on his paper, *The Proceedings of the Royal Society B* paper, "Genetic analysis of hair samples attributed to yeti, bigfoot and other anomalous primates," examined 30 samples from "museum and individual collections" that had been labeled as the North American bigfoot, Tibetan yeti, Mongolian almasty, and Sumatran orang pendek. According to a spokesperson for the journal, "The correction is being made because the institution the author Brian Sykes gave as part of his affiliation does not exist." The exact wording of the notice has yet to be decided.

According to Jonathan Leake at the *Sunday Times:* The paper gave Skyke's affiliation as the Institute of Human Genetics at Wolfson College, Oxford. Sykes is a fellow of Wolfson but he admitted the institute was mythical. "The journal required some sort of additional address in the college and, hey presto, I became an institute!"

This unfortunate affair has taken place embarrassingly soon after it transpired that the paper's claim that two samples were identical to those from an ancient polar bear is actually not true (see last issue). Oh dear.

SOURCE:
http://retractionwatch.com/2015/04/14/bigfoot-paper-corrected-because-it-doesnt-exist-the-authors-institution-that-is/

Mystery Cats

The return of the Missing Lynx

Following several sightings of a large cat resembling a lynx in spots around Newton Stewart and Minnigaff recently, the Lynx UK Trust has spoken of national public support for their reintroduction. Two runners using routes in Old Minnigaff are believed to have spotted unusually large creatures running off into the undergrowth over the last few weeks with one telling 'The Gazette' the sighting momentarily stunned him before he retreated and thought it best not to investigate. And such sightings could become more common after a public survey carried out last month by the Lynx UK Trust returned a remarkable 91% in favour of a trial reintroduction of lynx to the UK, with 84% believing it should begin within the next 12 months.

With forestry being the prime candidate for the project, Galloway Forest Park and other national parks across to the Borders could be used. But farmer, John Morren – who keeps cattle and sheep on his land near the Grumack Forest – said: "I think it would be negative. I think we already have badgers and foxes and they are all predators." The 78-year-old added: "I think it would be a bit alarming for the lambs to be picked up and carried and bad for the ewe when she loses a lamb. If a lynx was hungry it would go for the ewe as well."

Chief scientific adviser to the reintroduction project, Paul O'Donoghue, said lynx presented "very little threat" to livestock. Mr O'Donoghue said: "We've been blown away

by the level of interest and support from the public. That led to government approval for the trial reintroduction. "The UK public have spoken. Lynx have proven themselves across Europe to be absolutely harmless to humans and of very little threat to livestock, whilst bringing huge benefit to rural economies and the natural ecology." However, Huntly councillor Joanna Strathdee said: "I am not an expert but how can they be sure they are not going to kill farm livestock? I would question how they can be so sure they are not going to decimate the sheep and lambs."

Deputy director of policy for the National Farmers Union Scotland, Andrew Bauer claimed the Lynx UK Trust had "neither properly consulted land managers nor credibly explained how it plans to manage the risks" of the reintroduction. He added: "Lynx are solitary and territorial, with individual ranges of at least 40 square miles. It is difficult to reconcile this with Lynx UK Trust's proposals for reintroduction of multiple adults into relatively small forested areas." The trust will apply for licences for a controlled trial later this year.

SOURCES:

https://www.pressandjournal.co.uk/fp/news/aberdeenshire/562557/fears-for-aberdeenshire-livestock-as-lynx-reintroduction-plans-take-step-forward/

http://mysterycats.blogspot.co.uk/2015/01/newslink-north-wales-call-to-find-out.html

Aquatic Monsters

In A While Crocodile

The Metro carried this story in early February 2015: "A large unrecognisable object was spotted off the coast of Britain – so naturally people have started thinking a crocodile or some sort of monster is lurking in our waters.

Photographer Allan Jones took some pictures of the creature wallowing in the waters off Plymouth Sound in Devon.

He said the '20 foot long' animal was swimming against the current about half a mile offshore, and when he showed the images to a university technician he was told it was probably a crocodile.

Allan said: 'I've never seen anything like it – the first thing that struck me was that it looked just like a huge crocodile.

The creature or object moved in circles, appeared to curve its shape and moved a considerable distance from left to right, turn and then move back the other way.'

The paper speculated that it was an Indopacific crocodile, but we reckon that it is either a sunfish or a basking shark; two species well known from British waters, and are refraining from making sarcastic remarks about journalists with a lamentable lack of knowledge of Natural History because we are such nice people.

SOURCE: http://metro.co.uk/2015/02/09/massive-crocodile-or-maybe-a-monster-spotted-off-the-british-coast-5055424/

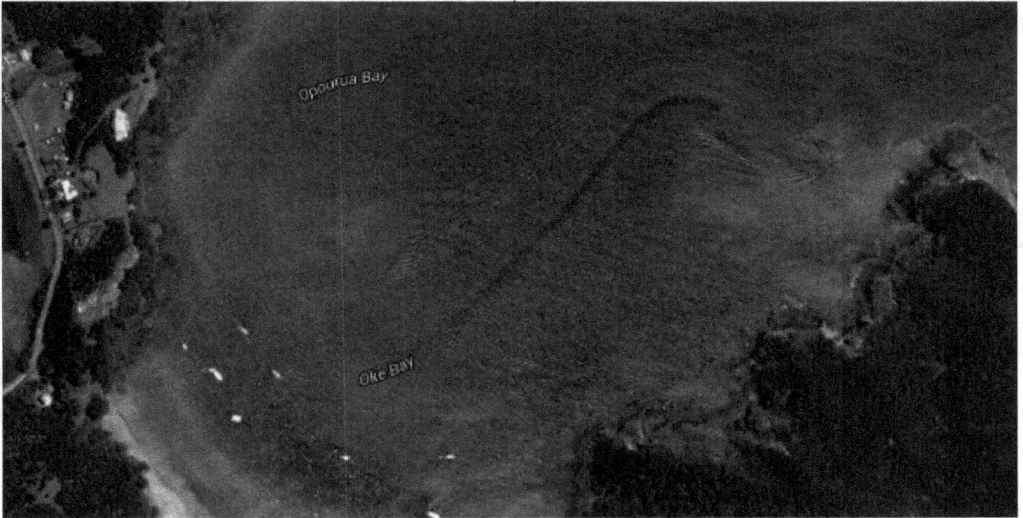

Another Google anomaly

Most people head to the Bay of Islands on the northern tip of the North Island of New Zealand for the big-game fishing and sailing. New Zealander Pita Witehira was looking at Google Earth photos of the area because he has property near Oke Bay. What he saw was not a game fish or a sailboat but the wake of a creature that appears to be 12 meters (39 feet) long.

Witehira, an engineer, was looking at a Google Earth satellite photo taken on January 30th when he spotted the mysterious and unusually large wake. He contacted Google and was told the company doesn't speculate on what anomalies in its photos might be but the person thought it was the wake of a boat.

Witehira disagreed with Google's suggestion because there is no white foam that would have been churned up by a boat motor. He had his own thoughts on what it might be. It's got to have a lot of weight under the water to create that kind of drag. The native Maori would call this a Taniwha as it appears not to be a whale and it is far too big to be a shark. It is moving too fast and turning too sharply to be a whale.

SOURCE: http://mysteriousuniverse.org/2014/12/sea-creature-spotted-in-new-zealand-waters-by-google-earth/

They needed an elephant gun

"It is true, it does exist, I saw it with my own eyes," says a Kelowna boater of a surprise experience with Ogopogo. Two women enjoying the sun and some fun time on the water had the scare of their life when they say Ogopogo swam near their boat. "It was a huge snake, I saw it, I saw the head. It was two feet thick and it was like 50-feet long. I could not believe it," says Suzie St-Cyr Cowley.

The "Ogopogo" was spotted about 1,000 feet off shore in front of Quails Gate Winery on the westside. She says it was heading south in the lake. "I could not believe it," says St-

Cyr Cowley. "I was afraid because we were so close and I wanted to move my boat away. I was screaming "Oh my god, that's Ogopogo! It was so big.""

"I thought it was just a story," says St-Cyr Cowley. "But, it is true, it does exist, I saw it with my own eyes."

Her friend, Marie Letourneau, had just arrived in town from Quebec for her first vacation in Kelowna and had never heard of Ogopogo. Letourneau was the first to see the creature and didn't know what she was seeing. "She was just yelling "Suzie, Suzie, there's a big thing", and I turned around and couldn't believe it."

She says because they were so in shock, and quite afraid it would go under the boat, they forgot to grab a photo or take video. "I yelled at her 'Did you take a picture?' but we both were so surprised," says St-Cyr Cowley.

"We could not believe our eyes. I turned around and saw the back and saw the head and it just went down. We are so sad we didn't take a picture." The women claim Ogie was very long and snake-like and had five curves along its back. "It does exist, I saw it with my own eyes. I see it and now I believe it," says an adamant St-Cyr Cowley. "I am sure he is around."

SOURCE:
http://cryptozoologynews.blogspot.co.uk/2015/04/boaters-say-they-saw-ogie.html

Another amusing promotional scam

So it is with that knowledge that a campaign has been launched to have Nessie recognised as the national animal of Scotland. Inverness cruise company

Loch Ness by Jacobite wants to replace the unicorn, a legacy from William I's decision to use the mythical creature on his coat of arms.

The first sighting of the Loch Ness Monster dates back to 565AD and the question of whether or not she exists is said to be worth millions each year to Scottish tourism.

SOURCE:
http://metro.co.uk/2015/04/23/loch-ness-monster-could-become-national-animal-of-scotland-5162923/

Carl Marshall's Column

It would thus appear that while, with very few exceptions, all so called 'Sea serpents' can be explained by reference to some well known animal or other natural object, there is still a residuum sufficient to prevent modern zoologists from denying the possibility that some creature may after all exist.

Encyclopedia Britannica
                    (Eleventh Edition).

It has been said that the sea serpent's name evokes negative connotations! This view is proven, conjuring up fantastic ideas of colossal sea snakes or even surviving Mesozoic reptiles such as *Plesiosaurus* and *Elasmosaurus,* by most scientists, journalists, and the man on the street alike; and that there is little more to this persistent phenomena than age old tales of dragons and mermaids. But is this habitual mode of thinking productive or even accurate?

Most scientists now accept the fact that large new species are still awaiting formal discovery in the dark depths of the worlds oceans, and discoveries of such unanticipated species as the giant squid *Architeuthis dux* (Steenstrup, 1857) the megamouth shark *Megachasma pelagios* (Taylor, Compagno & Struhsaker, 1983), and the Perrin's beaked whale *Mesoplodon hotaula* (Dalebout, Mead, Baker, Baker & van Helden - found in 1975 but formally described in 2002) should confirm this dictum in the minds of remaining skeptics. So why then, with a few exceptions such as a recent peer-reviewed technical paper by Dr. Darren Naish et al., is the sea serpent still confined mainly to Zoomythology and failed to find its way into conventional scientific journals?

An undiscovered sea serpent, or for that matter any other large bodied aquatic animal that has avoided detection, is not an absurd proposal when we take into consideration low population density coupled with cosmopolitan geographical distribution, unexpected biological functions and Parental investment (PI). However most explanations proposed to identify these animals are clearly nonsensical and this along with a continuing failure to procure a Holotype specimen has tainted the minds of the majority of scientists.

Proposed identity theories -

A true serpent.
Could not reports of long necked sea serpents be attributed to sightings of colossal snakes such as boas and pythons? This theory at first seems reasonable when we take into account some tropical species do reach immense lengths (approximately 30ft) and is probably correct in a few reports from tropical waters where large species, such as reticulated pythons *Python reticulatus*, do migrate between islands in shallow coastal waters, but what about sightings from temperate countries like Britain and a report from Yorkshire in 1934? As all living snakes are ectothermic (cold blooded - as are *all* living reptiles) this theory is extremely unlikely when we look at the problem as a global phenomenon and should probably look elsewhere for a more

Long Necked Sea Serpents: An overview of historical reports which might suggest evolutionary links to as yet undiscovered species in unexplored uncharted depths.

plausible explanation.

A prehistoric survivor.
The theory that long necked sea serpents are surviving long necked marine reptiles is arguably more appealing than the true serpent theory but still has serious inconsistencies to be addressed. Paleontological discoveries tell us species like *Plesiosaurus*, *Elasmosaurus* and their ilk were also probably ectothermic (gigantothermic) which, analogous to today's reptiles, would unable them to regulate their body temperatures in colder climatic conditions for prolonged periods of time by solely relying on the external heat from the sun. Therefore the idea that these huge bodied reptiles somehow survived the KT extinction event by it having less of an impact on marine ecosystems than terrestrial ones is unlikely, and considering this group has had no surviving contemporaries for 66 million years makes this all the more improbable.

The late University of Chicago Biologist Dr Roy P. Mackal once proposed the theory that long necked sea serpents and lake monsters are the result of surviving modern day zeuglodons (primitive prehistoric whales) which somehow survived extinction 37 million years ago.

Serpentine Cephalopods.
Henry Lee in his 1883 book *Sea Monsters Unmasked* proposed that longnecks might be the result of giant squids near the surface whose tentacles were elevated out of the water and could have been confused for elongated necks. If this theory is accurate, in my opinion I would suggest that these giant cephalopods are likely to be in their death throes as this activity suggests distress, probably from lack of oxygen due to the squid entering warmer surface temperatures, leading to suffocation and consequently the death of the animal.

Heuvelmans's Long necked Pinnipeds.
In his phenomenal book *In The Wake Of The Sea Serpents* Dr Bernard Heuvelmans helped promote a more convincing and realistic theory; that of his ten proposed aquatic sea serpent types, his long necked sea serpent, which he dubbed *Megalotaria longicollis*, and merhorse - *Halshipus olai-magni* are both hitherto undiscovered gigantic pinnipeds (seals) which, in addition, I theorise would spend much of their lives in the bathyal zone. The probability that the longnecks are more

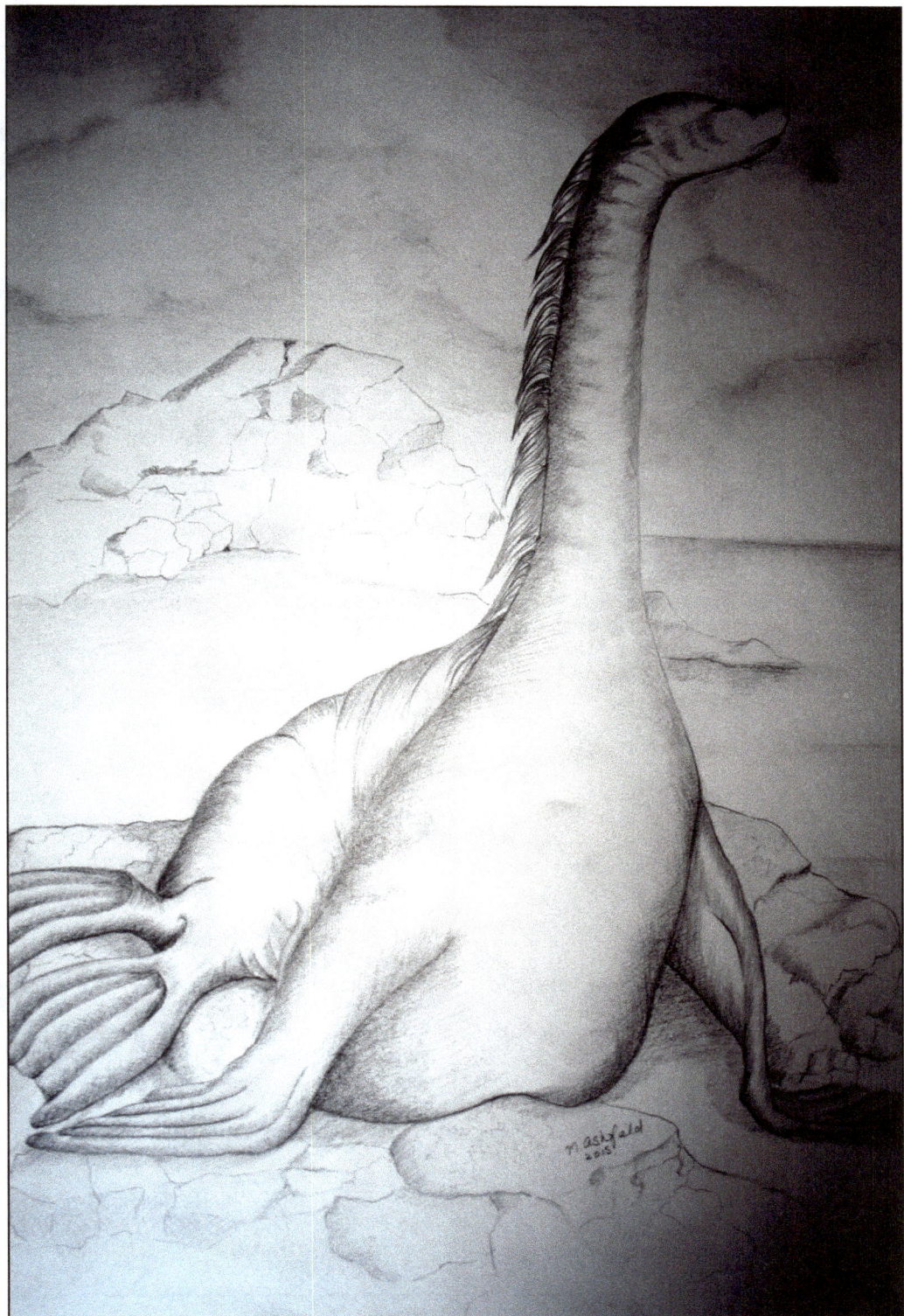

n ashfeld
2015

closely related to each other than to any extant pinnipeds seems more plausible to this author, and suggests that if formally identified and described, the creation of a totally new taxonomic sub-family, or maybe even one or possibly two totally new taxonomic families should be created to accommodate them. This and the proposal that longnecks probably birth at sea (or possibly in open bodies of water such as fresh water lakes) and hence would have short weaning and maturation times, which would be essential if they give birth at sea, makes them specialised enough to confirm this proposal.

Descriptions of Longnecks from Heuvelmans's research.
Long Necked Sea Serpent (48 sightings).

- Neck long or very long, at an angle to the head.
- A big median hump on the back, making it look like two or three humps close together.
- No tail, or a mere stump of a tail.
- Two horns [or ears] on the head (not always present).
- Mottled colour + long neck.
Fast speed (more than 13 knots).

Merhorse (37 sightings).

- Long floating mane.
- Slender medium length or long neck + only one dorsal curve.
- Very big eyes
- Long hairs or whiskers on the face
Heuvelmans's theory suggests that approximately 15ft to 100ft long, previously undiscovered plesiosaur-like pinnipeds might be behind many long necked sea serpent and lake monster sightings. This theory can be traced further back to Dutch biologist Dr Anthonid Cornelis Oudemans's comprehensive overview of the sea serpent phenomenon in his 1892 classic *The Great Sea Serpent.* Oudemans hypothesised that the majority of sea monster sightings could be defined by a 20ft to 200 ft cosmopolitan, long necked, long tailed and previously undiscovered species of

pinniped which he dubbed *Megaphias megaphias* (literally giant snake).

Our hypothetical long necked seal may even be traced further back to an illustration and description from 1751 published by James Parsons '*in diverse countries*'. However this description is rather vague and probably that of a fur seal.

The long necked seal theory seems more agreeable than the other proposed identities previously mentioned, being a pinniped that is thought to be a convergent representative of a hypothetical transition between seals and whales. Furthermore, if it is to be believed that temperate lake monster sightings are the product of modern day plesiosaurs, then it must be concluded that somehow they evolved and adapted a tolerance to varying climatic conditions in both fresh and salt water environments in which mammals are far more readily tolerant.

Another thing worth considering are the descriptions of the head and necks of these alleged animals. They are described as being able to hold their necks high out of the water, presumably viewing surface objects - this indicates they are maximally flexible through the vertical plane which is again a mammalian trait. In contrast, prehistoric plesiosaurs are now known to have possessed brittle necks that could not be held high out of the water in the swan-like position!

A possible ancestral contender.
Longer agile necks are common among pinnipeds and long extendable necks are a common trait of the Otaridae, however there are other pinniped species that also exhibit fairly long necks. The leopard seal *Hydrurga leptonyx* for example have fairly long necks which they use to quickly strike out at prey. There is even an extinct species of phocid seal dubbed the swan necked seal *Acrophoca longirostris* (Muizon, 1981) from the Miocene and Pliocene epochs (23.03 to 2.58 million years before present) in what is today the

Pacific coast of south America, which was very plesiosaur-like in outline and probably convergent with it - occupying the vacant ecological niche left behind by these extinct reptiles and could even be a direct ancestor to one or both of the cryptids with which this article is primarily concerned. Its neck appeared serpentine and was probably strong and agile enough to be held erect, enabling the seal a better view of objects on the surface. The swan-necked seal has been allied with the genus *Monachus* which includes such phocids as the critically endangered Mediterranean monk seal *M. monachus* and the now formally extinct Caribbean monk seal *M. tropicallis*.

In this article I wanted to bring to attention the highly elusive qualities of these famous aquatic cryptids, and focus on the possibility that these alleged animals may exist but can allude detection due to various factors such as the extreme depths in which they are said to inhabit, their obvious fluctuating population densities, (there seems to be a strong correlation between longneck sighting frequency and cod abundance data - Woodley, 2008) and their wide geographical distribution and shy nature. Although it has been proposed (Paxton, 1998) that most new species of large bodied aquatic animals will likely be cetaceans (whales and dolphins) and this is probably accurate. However one should not disregard the possibility of hitherto undiscovered long necked pinnipeds considering animals fitting this description have been consistently observed, often by multiple witnesses, for centuries.

It was unmistakably some kind of animal. The entire head and neck were covered in wet fur which lay close to the body and glistened in the afternoon sunlight. When it was almost beside our boat the head turned and looked squarely at us. My first thought was that we were seeing some kind of giant otter or seal, but I was immediately impressed by the fact that this was not the face of an otter or seal.

Thomas Helm in St Andrews bay off the

northwest coast of Florida in the Gulf of Mexico 1943.

Illustrations by Maureen Ashfield.
© Carl Marshall

References and suggested further reading.

Arment, C. (2004). *Cryptozoology: Science and Speculation.* Coachwhip Publications, PA, USA.
Buamann, E.D (1972). *The Loch Ness Monster.* (1st Ed) Biddles LTD.
Heuvelmans, B. (1968). *In the Wake of the Sea Serpents.* (1st Ed) Rupert Hart - Davis.
Kirk, J. (1998). *In the Domain of the Lake Monsters: The Search for the Denizens of the Deep.* (1st Ed). Key Porter Books, Toronto, Canada.
Mackal, R. P. (1976). *The Monsters of Loch Ness.* (Swallow Press, Chicago, USA.)
Oudemans, A. C. (1892). *The Great Sea Serpent.* Brill, Leiden.
Paxton, C. G. M. (1998). *A cumulative species description curve for large, open water marine animals. Journal of the Marine Biological Association of the United Kingdom.*
Riedman, M. (1991). *The Pinnipeds; Seals, Sea Lions and Walruses.* University of California Press, CA, USA.
Shuker, K. P. N. (2003). *The Beasts That Hide From Man: Seeking the Worlds Last Undiscovered Animals.* Paraview Press, New York, USA.
Thomas, L. (2014). *Weird Waters - The Lake and Sea Monsters of Scandinavia and the Baltic States.* (1st Ed) CFZ Press.
Woodley, M. A. (2008). *In The Wake Of Bernard Heuvelmans.* (1st Ed) CFZ Press.

Carl Marshall works at Stratford Butterfly Farm and is a fine field naturalist. Over the past couple of years he has become a very enthusiastic member of the CFZ, and his quasi-Fortean view of British natural history fits in perfectly with my own. He was, therefore, the perfect choice as a columnist for the brave new *Animals & Men*, and we are proud to have him aboard.

Looking for the Blue Devil

There's a saying in Belize. Don't call the alligator a big-mouth until you have crossed the river. I take it to mean, don't be so quick to speak until you are in a position to do so. When I was staying at a place called Black Rock in Belize last year (looking for arachnids, naturally) my 'guide' told me about a cave high up in the mountainside that had 'huge blue spiders' the size of your hand. The blue devil. The very notion demanded that the hairs on the back of my neck should stand erect. Locals do exaggerate sometimes, let's be fair. But Carlos is a very knowledgeable, competent and sensible lad who learned all he knows about his environment from living in it – not from books. A true auto-didact.

I asked if (when!) we could visit and he agreed. Now I do have a heart condition, and an implanted defibrillator and Carlos said that the climb was extremely difficult even for very fit people. Well you only live once and that immutable feeling of curiosity got the better of me. Better to die trying to find a magical blue spider than on the lavatory reading a chess book I say.

Carlos, myself and my intrepid wife Susan – a yampee Black Country girl - all set off the very next day. I almost died – or felt like I would at any moment. If my heart rate goes over 169 beats a minute I will receive an 'appropriate' 830 volt shock to try to get my heart in rhythm again so I

had to take it slowly, stopping each time my pulse went fast. It was almost vertical in most parts but I had to see the cave devil. It was a ludicrous climb for someone like me but they say the harder the struggle the more glorious the triumph and so it proved.

It was absolutely roasting hot but the cave entrance was cool and refreshing. Not many people came here, and I could understand why. As we stooped low to get through the initial labyrinth of nooks and crannies we were able to stand up. I switched on the torch and was overjoyed to see hundreds of bats (figure 1—opposite above) roosting overhead and in front of us. They didn't seem bothered by us and we just let them be. The guano was fairly thick underneath us and little creatures were living off it – nothing wasted in the natural world.

We spent a good deal of time scanning the walls and ceilings but no spider could be found. It was utterly exhilarating in the dark and quiet and we did find some large and extremely beautiful amblypigids scurrying around (figure 2—opposite below).

These are some of the weirdest creatures on the planet. Not spiders, not scorpions or crabs, they look like several arachnids were thrown into the air and this is what landed. In their own way though, they are extremely beautiful and no part of their design is wasted. Like a scene from a horror film, one touched my leg in the

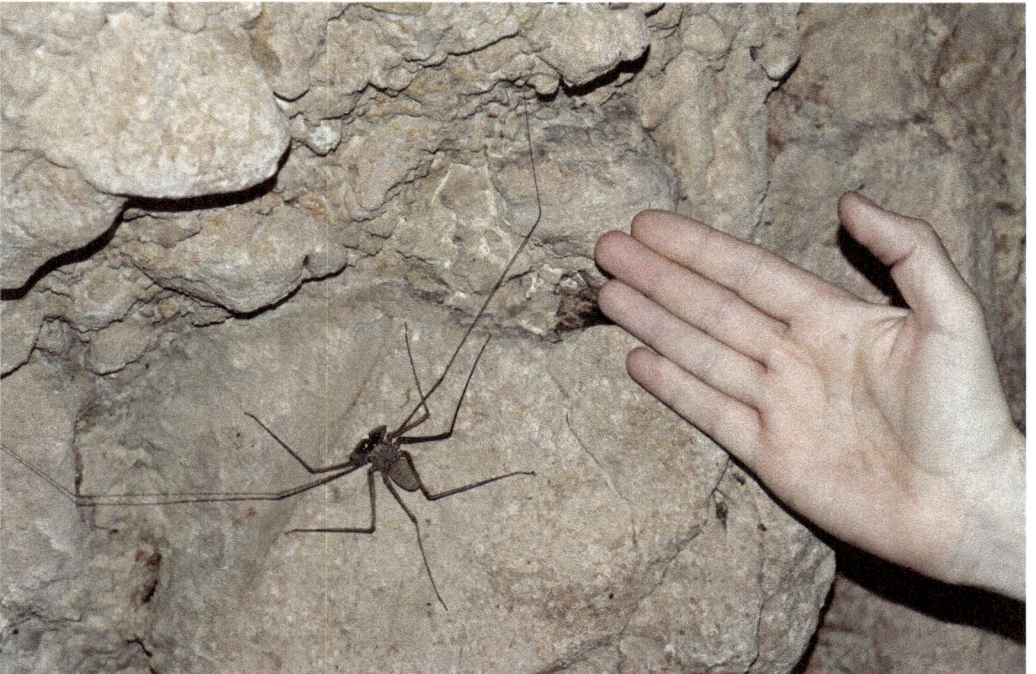

dark. My heart beat faster (always dangerous!) as I hope it would be that legendary blue spider. Alas and alack it was not.

We search deeper into the cave and Carlos stopped us in our tracks. "Oh my God" he exclaimed. "What the hell is it – the spider?" I enquired. It wasn't. Carlos had spotted a green/blue frog (figure 3—above) that he had seen before but he said was not yet described. It was indeed a wonderful little chap. This is what curiosity does for you. It reveals secrets, it creates magical moments.

On we went, then Carlos squealed like an excited child again. This time it was a lizard (Figure 4 opposite) and one that Carlos had only seen once before and only in the cave and again he could find no record of it in the books. How exciting. This creature had a super primitive head with a strange kind of helmet covering it. I am no expert on reptiles but I knew this was a special find. Again, curiosity gave us generous rewards. Deeper again we squeezed, bump and stumbled into the cave. We came across broken old pots that the Mayans had left. (Figure 5—below) The sense of history was beyond words.

Who had walked these caves before us? How did they live exactly? What were they thinking? What stories of superstition could they convey to me? I felt humbled and very respectful of the place. The Mayan spirits were definitely there. This was <u>their</u> cave and I was a guest. Would my respect pay off? Could they deliver a devil for me? It was ever so slightly scary but fear is the dark room where the devil develops his negatives. One must press on.

Suddenly there was a new danger – bees. Carlos (figure 6—opposite) was a brilliant guide but even he got it wrong in a hilarious way. Clearly we had stumbled into a bees nest and they were flying into the headtorch lights – and our faces. Susan yelped "I've been stung" then I got stung several times. "It's okay" Carlos said from a few yards away putting his hand towards the bees. "They are stingless". A moment later he jumped back like a cobra "Oh my God, I have been stung' he cried and the bees swarmed around him trying to inject the venom. Sue and I folded double with laughter. It wasn't so bad though. They did sting, but they were not so fierce. None of it mattered anyway, this is exploration and one is meant to endure such things or there would be no stories to tell, no bedtime memories, no moments of joy when we are old and grey and full of sleep. Right?

Nevertheless, the blue devil remained invisible. I had literally been up to my chin (lying horizontally photographing cave lice) in bat guano, stung by angry bees, cut on rocks and still there was no spider. Sadly it was time to leave. Carlos was clearly disappointed and apologised but we told him not to worry – this is nature and you take your chances. The blue devil will have to remain elusive but

everything we hoped it would be. I could have cried with joy.

I have no idea what species the spider is. It is probably some kind of wolf spider, or huntsman. For now, that matters not. I just know where it is, that it exists, that it is out there and that is enough for me.

Many a night I have gone to sleep thinking about the inside of that cave and the exquisite forms of life in the dark and the damp. There was no gold to be found, but there was a blue devil, and that dear reader, made all the difference.

it isn't the end, we'll be back. We steadily probed our way back through the maze, Carlos trudged just ahead of us. Hold on. Did I hear correctly? I cocked my head. "Quickly, get over here, I have found one" Carlos whispered as loudly as he could.

My heart raced again, but I made my way as fast as I could, not daring to believe it could really be the Golden Fleece. Perhaps it was something else. Please let it be the spider, please dear Mayan spirits, help us in our hour of need.

And there it was. Glorious, majestic, effulgent...BLUE! This was not a tarantula (theraphosid) but a true spider of some kind. Apparently it only lived in the cave.

I managed to get a few shots with my shaking hands before it disappeared suddenly into a crevice, back to the spirit world. True, it was blue and it was the size of my hand. (figure 7—on p.36) I could not put anything by it to get scale as it would have made off, but trust me...it was

Born in Birmingham, England Carl Portman has always followed the maxim 'interest is where you find it' and this certainly applies to natural history. He has bred endangered species of tarantula spiders, written two books on natural history travel, and lectures around middle England on animals and rainforests. Oddly he has a diploma in sexing juvenile theraphosid spiders, is an English Chess Federation County Chess Master, supports Aston Villa and has a strange addiction to Turkish Delight (covered in chocolate). Having worked for the Ministry of Defence for 30 years he now spends his time doing lecturing, chess coaching, some photography and management consulting.

He has spent time studying animals in the rainforests of Australia, Ecuador and Costa Rica searching for new and ever curious insects and arachnids and has a desire to find a new species somewhere in the world. His motto is 'Don't complain about the dark, light a few candles'.

He is married to Susan and lives in Oxfordshire. Their two Border collies, Darwin and Dickens keep them fit and ensure that there is never a dull moment in the household.

watcher of the skies

CORINNA DOWNES

In another period where bird poisoning has once again been rife, for example the 13th white-tailed eagle in Ireland being found dead in her nest in April, and in South Africa, 200 of its national bird, the blue crane, being poisoned allegedly by a farmer, Malta also rejected a spring hunting ban, meaning that the spring hunting of quail and turtle doves, an activity that is banned across Europe, will continue unabated in Malta after voters in a referendum rejected a proposed ban. It was decided on a razor thin margin, with just 2,220 more votes deciding against the ban out of a total of 250,648 votes cast. The "yes" camp won 50.4% of the vote thanks to a strong showing for the pro-hunting contingent on the island of Gozo, which is part of Malta. BirdLife Malta, which led the campaign against spring hunting, described it as "missed opportunity to end the killing of birds in spring".

However, news of the first BTO-tracked cuckoo returning to Britain from Africa brought some light relief. Although it had been thought that the bird called Dudley would win, an outsider called Hennah (named after First Lieutenant William Hennah who was on HMS Mars in the Battle of Trafalgar) came back first.

Further details can be found at www.bto.org/cuckoos. http://www.wildlifeextra.com/go/news/bto-cuckoo-race.html

Unusual bird spotted at Upton Warren en route to Eastern Europe

An unusual bird – a red-necked grebe, complete with its summer breeding plumage - dropped in to Upton Warren Nature Reserve on its way to Scandinavia or Eastern Europe to breed in April. The reserve is between Droitwich and Bromsgrove, and John Belsey, a volunteer warden there, said: "A small number of red-necked grebes overwinter in the UK and they pass through on their way to and from breeding grounds in spring and autumn. This is the first time we've ever seen one at Upton Warren and to have one in full summer breeding plumage is absolutely fantastic. On Wednesday there was only one other red-necked grebe being

reported in the UK so we're really honoured – and lucky – that this one chose Upton Warren."

SOURCE: http://www.eveshamjournal.co.uk/news/regional/12915374.Unusual_bird_spotted_at_Upton_Warren_en_route_to_Eastern_Europe/

Bryan's Shearwater is confirmed breeding in Japan's Ogasawara Islands

The breeding site of the recently described critically endangered Bryan's shearwater (*Puffinus bryani*) has been suspected to be on Japan's Ogasawara Islands where corpses of birds have been previously found. Breeding has now been confirmed with an incubating bird being discovered on Higashijima Island in the Ogasawaras, as described below by the Mainichi Japan of 25 March.

"A team of scientists has confirmed a nesting site of an endangered seabird species once thought to have gone extinct on the Ogasawara island chain, it has been learned - - the first time a nesting site of the species has ever been discovered." The species, whose body length ranges between 27 and 30 centimetres, was believed to have gone extinct after it was last seen on Midway Atoll in 1991. Scientists conducted DNA testing on seabirds found on the Ogasawara Islands between 1997 and 2011, as their features matched those of the Bryan's Shearwater. In 2012, it was confirmed that the birds were indeed members of the Bryan's Shearwater species. The Ministry of the Environment subsequently included the birds in the Red List as a critically endangered "IA" species. The researchers

spotted a flock of 10 Bryan's shearwaters on Higashijima Island, approximately three kilometres east of Chichijima Island, on Feb. 25-26. One of those birds was holding eggs inside of its nest.

SOURCE: http://www.rarebirdalert.co.uk/v2/Content/ ACAP_Bryans_Shearwater_confirmed_breeding_in_J apans_Ogasawara_Islands.aspx?s_id=732063626

'Extinct' bird rediscovered: Last seen in 1941

A scientific team has rediscovered a bird previously thought to be extinct in Myanmar. Jerdon's babbler (*Chrysomma*

altirostre) had not been seen in the country since July 1941, where it was last found in grasslands near the town of Myitkyo, Bago Region near the Sittaung River.

The team found the bird on 30 May 2014 while surveying a site around an abandoned agricultural station that still contained some grassland habitat. After hearing the bird's distinct call, the scientists played back a recording and were rewarded with the sighting of an adult Jerdon's babbler.

The small brown bird, about the size of a house sparrow, was initially described by British naturalist T. C. Jerdon in January 1862, who found it in grassy plains near Thayetmyo.

SOURCES:
http://orientalbirdclub.org/2015/03/05/jerdons-babbler-rediscovered-in-myanmar/

http://www.sciencedaily.com/ releases/2015/03/150305110237.htm

New species of tapaculo in South America

After being misidentified and sitting in a museum drawer in the Smithsonian for more than seventy years, a group of bird specimens collected in Colombia and Venezuela has been determined to represent a previously unknown species, now dubbed the Perijá Tapaculo (*Scytalopus perijanus*).

In a new paper in *The Auk: Ornithological Advances*, Jorge Avendaño of the Universidad de los Llanos and his colleagues describe how the Perijá Tapaculo differs from the other birds in its genus in its genetics, appearance, ecology, and vocalizations.

Tapaculos are a family of mostly small black

or brown songbirds found in South and Central America, which forage for insects in grasslands and forest undergrowth. In 1941 and 1942, ornithologist Melbourne Carriker, Jr. explored the western slope of the Serranía de Perijá mountain range on the Colombian-Venezuelan border, where he collected 27 tapaculo specimens and sent them to the Smithsonian. At the time, they were mistakenly identified as _Scytalopus atratus nigricans_, a similar bird found at lower elevations, and in the following decades these upper montane tapaculos remained overlooked and unstudied.

However, in 2008 and 2009, a new set of specimens and sound recordings was collected in the same region visited by Carriker, and Dr. Avendaño and his colleagues were able to conduct a genetic analysis as well as analyzing the birds' appearance and calls. The newly-named Perijá tapaculo is a small bird with a buffy belly, gray back, and brown nape, and its song and calls are distinctly different from those of other tapaculos.

Unfortunately, due to habitat loss in the region, this new species is already in trouble. "The species is more seriously threatened on the Colombian slope than on the Venezuelan slope because its habitat is protected by a large National Park in Venezuela," explains

Dr. Avendaño. "So, we recommend establishing a new national park or a network of reserves in Colombia connected to the Sierra de Perijá National Park of Venezuela. This binational park is necessary in order to effectively preserve the high species diversity and endemism of birds and other biological groups of the region."

SOURCE: http://www.sciencedaily.com/releases/2015/03/150311160531.htm

Hummingbird, thought extinct, rediscovered in Colombia

When conservationists Carlos Julio Rojas and Christian Vasquez went into a Columbian mountain range looking to document fires burning in the fragile ecosystem, they ended up rediscovering a hummingbird that had not been seen since 1946 and was believed to have gone extinct. The pair were able to take the only known photographs of the blue-bearded helmetcrest.

Rojas said: "I managed to take a quick photo of it before it flew off. I then reviewed the photo on the camera screen and immediately recognized the strikingly patterned hummingbird as the long-lost blue-bearded helmetcrest. I was ecstatic. After reports of searches by ornithologists failing to find this spectacular species, Christian and I were the first people alive to see it for real." The bird is known for its prominent crest and elongated throat feathers forming what looks like a beard, according to Birdlife International. In the centre of the beard are metallic purplish-blue feathers, and the tail has an extensive buff-white area.

Found only the Santa Marta region of northeast Colombia, the bird had seen its numbers drop due to habitat loss caused by deforestation, overgrazing from herds of cattle and extensive burning by the Kogi indigenous people for farming. The species is dependent on stunted forest and bushes amongst natural páramo grasslands - habitat that is highly susceptible to fires during the dry season. The situation is even more difficult because the flowering plant the helmetcrest depends on - the Santa Marta Frailejon (*Libanothamnus occultus*) - is itself threatened by persistent fires and has also been declared critically endangered. "Sadly the survival of the blue-bearded helmetcrest hangs by a thread," Rojas said.

SOURCE: http://www.cbsnews.com/news/hummingbird-rediscovered-after-nearly-70-years-in-colombia/

Fig 1 CAMPYLOPTERUS PHAINOPEPLUS
2 OXYPOGON CYANOLÆMUS.

1,000 twitchers in a flap as 'lost' rare bird flies into Britain

Late April brought a rare Hudsonian godwit to the Somerset Levels. This bird usually migrates to South America in the winter and then heads north to its breeding grounds in Canada and Alaska in spring and has only been spotted in the UK a handful of times.

One birdwatcher, Michael Trew, 70, said: "It is quite a strange affair. We don't know how long it has even been here. It's not supposed to be here at all."

It is thought that the bird must have got confused and accidentally followed a flock of migrating birds across the Atlantic.

A Hudsonian godwit has been recorded in 1988 in north-east Scotland, with another one popping up in east Yorkshire and in Devon three times between 1981 and 1983.

The bird, which has long dark legs and a long bill, has joined its English cousins, the black-tailed godwit at the Shapwick Heath Reserve near Glastonbury.

SOURCE: http://www.express.co.uk/news/nature/573061/Rare-Hudsonian-godwit-bird-unexpectedly-flies-UK-birdwatching

Great Blue Heron

The 14th April brought with it the discovery of Britain's second great blue heron on St Mary's, Scilly. The heron's occurrence is surrounded by amazing coincidences: not only is it on the same island as Britain's only other record, but was also found by the same observer! It is a common bird near the shores of open water and in wetlands over most of North America and Central America as well as the Caribbean and the Galápagos Islands. It is a rare vagrant to Europe, with records from Spain, the Azores, England and the Netherlands.

SOURCE: http://www.birdguides.com/webzine/article.asp?a=4963

Influx of Eurasian Hoopoe

During the middle of April there was an influx of Eurasian hoopoe, the event being unlike any other this century. There were no fewer than 168 reports of the species, although - in reality - hundreds must have arrived. According to Birdwatch Ireland's head of operations, Oran O'Sullivan, "It is 50 years since so many hoopoes have been spotted here. Usually, fewer than 10 are recorded in early spring or late autumn when migrating birds stray off course. He added that the exotic birds, about the size of a starling or thrush, were a Mediterranean species, typically nesting in trees and olive groves.

"They have very big wings and when they take off you see a flash of black and white. When they land they throw up this crest, like an Indian chief's head dress. They are exotic all the way." He said the birds wintered in Africa and could fly as far as northern France. "Even a few breed in the very far south of England. They come up in good weather and in spring they can overshoot France and hit Wexford."

SOURCES:
http://www.birdguides.com/webzine/article.asp?a=4963

http://www.irishtimes.com/news/environment/hoopoe-causing-a-hoopla-in-southeast-as-50-exotic-birds-spotted-1.2190020

Last of Lady Amherst's Pheasant? Twitchers flock to see one remaining Lady Amherst's pheasant in wild for last time

Only one specimen of the dazzling Lady Amherst's pheasant remains in the wild in Britain in a top secret haunt and devoted twitchers are hurriedly trying to track it down.

The pheasant is named in honour of Sarah, Countess Amherst, who was married to William Pitt Amherst, Governor General of Bengal.

Lord Amherst sent the first specimen back

to London in 1828, although the bird did not survive the journey.

Josh Jones, of the BirdGuides information service, explained the reasons for the pheasant's allure: "Though introduced species may not be to everyone's taste, there can be no argument over the beauty of Lady Amherst's pheasant. One of the most stunning yet most elusive species on the British list, it has traditionally been considered highly desirable by birders and its ever-increasing rarity only adds to its enigmatic aura. If it proves that this male is indeed the last of its kind surviving on the Greensand Ridge, it represents the sole legacy of a population which has enchanted British birders and drawn them to this pocket of the Home Counties for a number of decades, something of a Holy Grail. Once this final individual perishes, it will bring a colourful era to an end."

While other birds that have been threatened with extinction in Britain, such as the red kite, white-tailed eagle and chough, have made a comeback to former haunts through successful re-introduction programmes, there are no such plans for Lady Amherst's pheasant because under the Wildlife and Countryside Act non-native species cannot be released into the wild.

The Lady Amherst's pheasant was intentionally introduced at Woburn in the 1890s, and the main British population established itself along the Greensand Ridge of Bedfordshire, spreading naturally west into Buckinghamshire. The extensive mixed pine, ash, oak and beech woodlands in the area (with an introduced rhododendron understorey) appear to have made a passable analogue for its native deciduous forest and bamboo thickets. However, similar habitat elsewhere in

Britain seems to have been unable to sustain numerous other introduction attempts over the last two centuries.

The bird is native to south-western China and Burma, and was added to the British Ornithological Union's British List in 1971 on the basis of the Bedfordshire and Buckinghamshire population, which by then was estimated to have reached up to 200 pairs during surveys at the time for the British Trust for Ornithology's *The Atlas of Breeding Birds in Britain and Ireland*.

SOURCES:
http://www.birdguides.com/webzine/article.asp?a=4956

http://www.express.co.uk/news/nature/569852/Birdwatchers-flock-see-last-Lady-Amherst-pheasant-Britain

Camera Traps Catch Rare Amazon Bird Following Peccaries

The rufous-vented ground cuckoo (*Neomorphus geoffroyi*) is currently listed as Vulnerable by the IUCN Red List and scientists know almost nothing about the bird.

Renzo Piana, the director of science and research with the Amazon Conservation Association, described the bird as "rare," "cryptic," "mainly solitary," and "mostly silent" - much of which explains why so little is known about it.

But camera traps are helping to reveal more about this, and thousands of other little known species. Piana and colleagues recently documented never-before-seen behaviour of the rufous-vented ground cuckoo on a camera trap in the Peruvian Amazon.

NEOMORPHUS SALVINI

A series of photos shows the cuckoo boldly following a group of collared peccaries (*Pecari tajacu*).

Piana explained: "It is suspected that the cuckoo benefit by increasing their chances of finding food," and added that the cuckoo probably scavenge peccary leftovers as well as insects on the run from peccary herds that can reach as many as 50 individuals.

Piana went on to say, "It is suspected that the species relies on primary forest. Major threats are habitat loss due destruction of primary forest for agriculture and cattle ranching, and road construction that leads to forest fragmentation."

SOURCES: http://news.mongabay.com/2015/0421-hance-roufus-vented-ground-cuckoo.html#ixzz3YQWOngvx

THEY COME OUT AT NIGHT
WWW.NIGHTBEASTS.COM
NIGHTBEASTS
TREEHOUSE PICTURES PRESENTS "NIGHTBEASTS" ZACH GALLIGAN AUDRA WISE
APESANAHQUAT SONNY SKYHAWK ROBERT MIANO BILLY DAYDODGE
HOLLY WILSON CHAD TRAGER DONN ANGELOS

INTERVIEW WITH WES SULLIVAN WRITER AND DIRECTOR OF NIGHTBEASTS

There is a long and venerable tradition of forteana in film. Some of the very earliest works from Georges Melies featured fortean subjects with films like *A Trip to the Moon* and *The Impossible Voyage.* The bond has continued covering all areas of fortean interest, *The Giant Behemoth* (sea serpents), *The Abominable Snowman* (yeti), *Earth vs. The Flying Saucers* (UFOS) and *The Haunting* (ghosts).

One of the latest on the list is *Nightbeasts* wherein a tribe of angry sasquatch come into conflict with humans. It comes from a venerable line of bigfoot horrors that started way back with *The Legend of Boggy Creek* in 1972. A low budget movie, not tied down to the creativity sapping shackles of big studios, *Nightbeasts* promises something that many mainstream Hollywood movies don't, originality.

The writer and director Wes Sullivan turned out to be quite a fan of all things fortean as well as classic horror movies. He was good enough to give me an exclusive interview about the making of the movie and his thoughts on bigfoot and the film industry.

RF: Many thanks for agreeing to be interviewed. I think first of all I'd like to ask you how you got started in the film business?

WS: No problem, it's a pleasure to talk with you.

I got started in the film business a long, long time ago while still an undergrad in film school. I first worked in traditional animation as a cel painter. It is a process that has since gone digital like so much else. Back in the old days every animation drawing was meticulously traced onto clear acetate animation cels using ink and a drawing pen. It was usually women that did this as they had steadier hands than men. Once the drawing had been inked onto the cel it was turned over on its reverse side and I and about 12 others would use a sable brush and paint a special paint made for animation called cel vinyl. This special polymer was made by a company in Culver City CA called Cartoon Color. I believe in Britain you have a company that still makes an almost identical product called Chomacolour. The painting of the cells involved having a certain skill. It couldn't be applied in too gloopy a fashion or too wet a fashion. It had to be laid flat and you had to be relatively fast. The reason being, a single character might have 7 or 8 colors. That meant cleaning your brush and having 7 or 8 small cups of paint to dip into and apply. You also couldn't paint a new color until an already applied color had dried. Also, a 30sec commercial might contain 720 of these painted cels or more. Some of the

Richard Freeman

commercials I worked on were Kellogg's Tony the Tiger, Rice Crispies Snap, Crackle and Pop, and the Trix rabbit. I also worked on educational films eventually moving up to become animator. It's important to know that I did all this while majoring in live action film and video in College. I wanted to be a director even then.

RF: Is this your first horror feature?

WS: Yes this is my first horror feature. After nearly 20yrs working in animation I got to realize my dream of directing a live action feature film.

RF: The film has yet to be released in the UK. Can you give us an outline, without spoiling the plot?

WS: *Nightbeasts* is the story of a recently divorced father, Charles Thomas who is brilliantly played by actor Zach Galligan who was recently in *Hatchet 3* but known best for the Gremlins movies. Charles Thomas (Galligan) takes his son Tim on a weekend hunting trip in an attempt to bond with his son but that trip is violently interrupted by the re-emergence of Sasquatch like creatures we call Nightbeasts, hence the movie title.

RF: There is a whole sub-genre of ape-man on the rampage in movies including *Snowbeast*, *The Creature of Shadow Lake* and *Night of the Demon*. But it all started with Charles B Pierce's seminal *The Legend of Boggy Creek*. I always felt that the low budget and documentary feel of that film makes it an outstanding piece of cinema. Did the film influence you in any way?

WS: Funny that you would mention Charles Pierce's *The Legend of Boggy Creek*. I remember that film very clearly because it was probably the very first scary movie that my Mom took me to see in a movie theatre when I was a kid. My Mom was always taking me to see scary movies that I shouldn't be seeing because I was too young but she didn't care. This is one of the things that makes my Mom a really cool person but that's another story.

To answer your question, yes. It was a major influence but not just in the making of *NIGHTBEASTS* but in the formation of my having an inquisitive and speculative personality. It put me on a definite path of wondering about the great mysteries of the Earth and the Universe. That film was the lodestone, the catalyst for my fascination with things like the missing colony of Roanoak, the lost land of Lemuria, the Alexandrian library lost to us in antiquity, the Himalayan Yeti, the Abominable Snowman, Orang Pendek, remnants of Noah's Ark in Mt Ararat Turkey, Atlantis, Shadow People, Roswell New Mexico, Spontaneous Combustion, Ball Lightning, Lizard men, Mokele Mbembe and the list goes on and on.

The other great influence on this movie specifically would have to be Hammer productions *The Abominable Snowman* starring Peter Cushing and ably directed by Val Guest and written by the inimitable

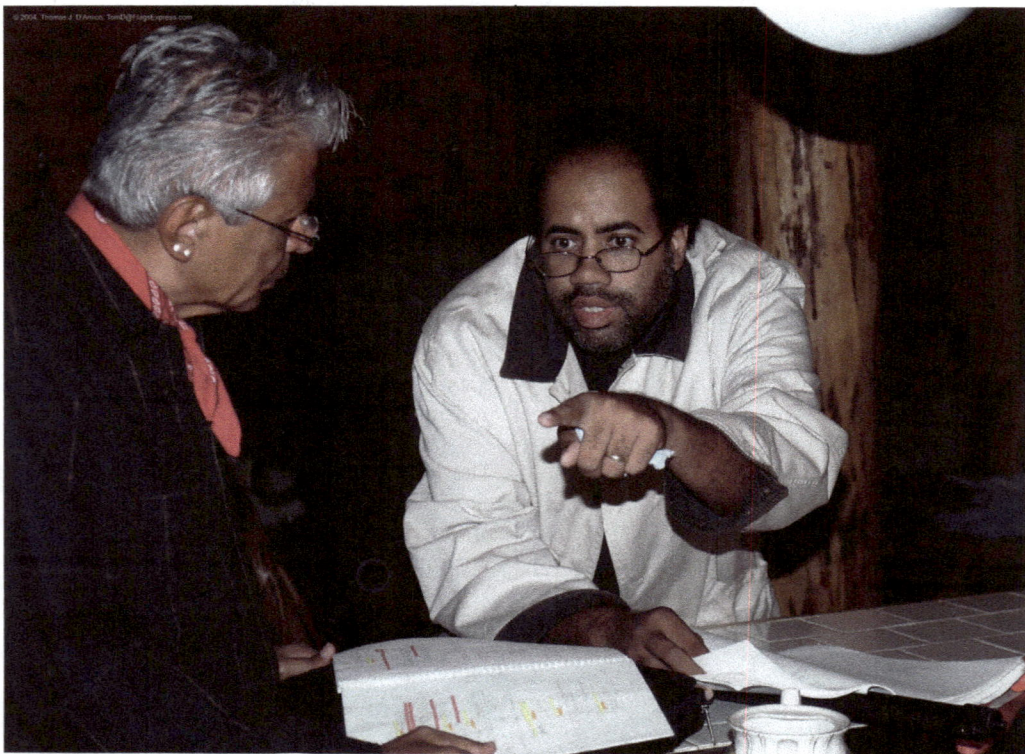

British god of Sci-fi Nigel Kneale whose *Quatermass and the Pit* still haunts my dreams many years after seeing it.

RF: Was there a particular event, report or case that inspired you when writing the script?

WS: Certain events in the script for *NIGHTBEASTS* were amalgamated from a number of eyewitness encounters with Bigfoot that are recorded in various newspaper articles, 911 calls, periodicals and documentaries that made up the comprehensive research that I did on the subject over quite a number of years.

RF: Were there any hairy moments, if you

will excuse the pun, during filming?

WS: There were a number of hairy moments that occurred during the production of the movie. One that is quite interesting involved a scene in which a man is attacked by a creature in the presence of his dog. Now the movie dog was very fond of the actor with whom he was doing the scene and when the time came to have a stunt performer (dressed as a Sasquatch) to attack, this normally very calm dog reacted out of pure instinct to protect the actor and started violently barking and eventually attacked the stunt performer, jumping on him and biting. It took 4 crew people to pull the dog off. We were forced to stop filming. The stunt performer removed his kit to show the dog that it was merely play acting and that he was human. The dog calmed down and relaxed. When we resumed shooting the scene and did Take 2 the exact same thing happened again. It happened another 3 times until we finally had to remove the dog from the set to film the scene. Luckily, the stunt performers costume was thick enough that he didn't suffer any injury.

Much of the movie was filmed in a small California town at an 8000 ft. altitude in a Pine forest that was hundreds of years old. The only access road to this location was a very narrow and very winding road that was a 45 min drive from the base of the mountain to the top where our location was. The road had only 2 lanes and many of the turns were blind and at times your vehicle was mere inches from a drop of several thousand feet, certain death. As a result, one of the actors (who was afraid of heights) routinely drove in the middle of the road on top of the solid yellow line up and down the mountain. Well, on one of these occasions, a truck was coming down the road in the opposite direction that he didn't see until seconds before an imminent collision and he was forced to swerve out of the way. He missed hitting the truck but started skidding off the road and into a dirt turnout at the very edge of the road. One of his tires hovered over the chasm below. If the turnout ditch wasn't there he'd be dead. It was as close a call as you can get. I personally hated driving that road which I did nearly a dozen times if you include location scouting.

RF: In reality there are very few accounts of attacks on people by sasquatch or other mystery apes. In the twenties and thirties gorillas were portrayed as savage monsters on screen despite them being generally unaggressive in real life. Nowadays the same seems true of bigfoot. Most movies tend to show it as a dangerous killer (a notable exception being Hammer's 1957 film *Abominable Snowmen*, based on a screen play by Quatermass creator Nigel Kneal). Why do you think this is? Did it bother you to show an essentially gentle creature as aggressive?

WS: I know exactly why it is. It has to do with the laws of drama which haven't changed since the times of the ancient Greeks. All good stories or narratives are based on conflict. i.e. Man vs. Man, Man vs. Nature, and Man vs. Himself. I'm proud to say that our story in *NIGHTBEASTS* embodies all 3 tenents of drama to a certain extent. Look at any story told since time immemorial , be it theatre, a novel, comic books, movies or oral storytelling tradition and you will find one or more of these tenets. Bigfoot in movies is usually manifest as the Man vs. Nature narrative.

Knowing what I know from my research about the Sasquatch and its peaceful if not benign nature, I would be lying to you if I said I didn't struggle with making Bigfoot the antagonist in the movie. That is until I found a way to rationalize it as a writer in the narrative sense.

We treat the creatures in the movie world of *NIGHTBEASTS* as just ONE of a number of sub-species of apelike creatures as there are in reality. I made no attempt to depict the "definitive Bigfoot" this is just OUR Bigfoot. As such we portray them as a particular sub -species that is "re-emerging" after a very long period of slumber. I took a page from real life and what is happening in the Amazon with the rise of viruses like Ebola and the MRA bug because of human beings settling in areas that were previously rain forest and as a result the natural balance of the ecosystem is thrown off-kilter. Our

NIGHTBEASTS have been asleep for several hundred years, a hibernation period. The town that our protagonists live in was their natural territory way back when. Hence our creatures are simply doing what any animal would do in claiming their territory. This does not make them truly EVIL. But humans are in their way and as a result there is conflict.

RF: Do you think such a creature as bigfoot / yeti could exist in remote parts of the world such as Tibet or the Rocky Mountains? If so what do you think it is, an unknown ape or an off shoot of one of mankind's ancestors?

WS: Yes I truly believe that these Bigfoot creatures exist in the real world and have existed for millennia or longer in remote places. They are apparently smart enough to avoid us. You can quote me on that. There I said it. Stories about space flight were once the realm of Science Fiction and now it's Science Fact. The same may be true for Bigfoot or Sasquatch stories in the future. In recent history the mountain gorilla was thought to be merely myth. Now we know better. Sasquatch may very well be an unknown ape but he also might be a distant relative of homo-sapiens who can say. If revered scientists like Jane Goodall and Jeffrey Meldrum can entertain the notion of Bigfoot's possible existence, I certainly can.

RF: Ape-man movies feature the North American bigfoot with very few using creatures reported from other parts of the world. Do you think this is simply a budgetary restriction?

WS: This is certainly due to issues of budget. Unlike other art forms like writing, painting, sculpture etc. filmmaking is a very expensive endeavour. It very well might be THE MOST EXPENSIVE of all the art forms. Even an independent non-studio movie like NIGHTBEASTS is expensive to make. It took me years to raise the money from private investors. The biggest expense of any movie after counting the above the line costs like "stars" is the "labor". It is the physical making of the movie that is costly, after that it is locations. We had twenty different location moves in our movie. That is a lot for an independent film. A movie is everytime the crew, lights camera etc. is set up in a new place hundreds of miles away from the last. That involves fuel and transportation rental. It is significant. That is why many movies are set in a single location. As far as Bigfoot movies go, NIGHTBEASTS is pretty epic in scale and its photography which is widescreen Techniscope.

This movie was shot and originated on Kodak 35mm motion picture film. It was not shot digitally. Digital is not good at capturing landscape and nature's grandeur. Contrary to the propaganda, film is not expensive. Labor is expensive.

For a movie about Yeti or Orang Pendek to travel around the world and have exotic indigenous locations it would need a budget 100 times larger than ours. And a film of that scale would have to be done with big Hollywood studio resources. Now we're talking about 20th Century Fox, Disney or Warner Brothers for that.

RF: There have been a few other cryptozoological creatures in movies like giant anacondas on Anaconda, the Mongolian deathworm in the eponymous flick and the Tasmanian wolf in Hunter and Dying Breed but largely it seems like an untapped field. Monster movies these days just seem to repeat zombies, werewolves and vampires endlessly in one long boring parade. Would you like to use another cryptozoological creature in one of your films/ If so which one?

WS: Hollywood will keep making Zombies, Werewolves, and Vampires until the cows come home because the industry is largely bereft of an ounce of originality. The corporations that run the industry are terrified to take a chance on something original hence the endless remakes and adaptations of young adult fiction that's been best sellers. I live in Hollywood but I AM NOT HOLLYWOOD so yes, I've written a treatment for a film about the Lizard Men that I will write as a Screenplay and try to make.

RF: What are you working on at the moment and can you tell us about any future projects?

WS: I'm currently in pre-production on a Sci-Fi movie involving bad and good aliens that is sure to be a lot of fun but I can't talk much about it right now.

RF: Finally what advice would you give to people interested in getting into writing scripts and screenplays?

WS: My advice to people wanting to write scripts or screenplays is to NOT do it. That might sound shocking coming from me a writer/director but I'm only half joking. Except for a few exceptions, the vertical corporate integration of the movie studios in the 21st century means there is an emphasis on franchises and brands. This means superhero movies, big Hollywood tentpole action thrillers with big named stars and sequels or remakes of previous movies that have brand awareness. Yes, the marketing maniacs are running the movie asylum and they have free range. Fewer films are being made by Hollywood and the ones that are have increasingly bigger budgets in excess of 100 million dollars. Oh yeah, and they have lots of CGI (Computer Generated Imagery)

Now I have nothing against these kinds of films in and of themselves. They are what they are .But what I do have against this is when these are made to the EXCLUSION of all else and THAT IS WHAT IS HAPPENING. The medium and smaller budget movies the "B films" are being eschewed in favor of the large tentpole releases. So where does the " original" screenplay fit into all this that is made whole cloth from a talented writer's fertile imagination. IT DOESN'T. Television is a far friendly medium to the writer these days. A writer is probably better off penning a fantastic story as a NOVEL and having it become a bestseller that he can then option to Hollywood for a king's ransom (That is if it appeals to that young adult genre. Think Harry Potter)

Many writers earn their keep doing re-writes or polishes on the big budget tentpole movies. This can be very lucrative work. Many of these writers buy large mansions on the hill that they live in. But the work is not so "creative". Meanwhile creative hypenates ,writer/directors like myself slug it out to realize our dreams, raising money, filming on modest budgets and reaching our core audience. In this there is much creative satisfaction. This is why I do this, not to live in a mansion because I can't afford one.

RF: Many thanks for your time.

WS: I've enjoyed this interview. Thanks it's been fun. Cheers

The changing face of the species list of British Lepidoptera

Jonathan Downes

There is a popular fallacy within cryptozoology that every cryptid is a 'prehistoric survivor' that has somehow avoided discovery and which lurks somewhere in the wilder areas of the globe. This is just simply not true, and works using two completely fallacious principals.

1. Evolution somehow stopped when we reached the modern era
2. The fauna of a specific biotope is static until somehow the predations of humankind wear it down.

Both these assertions are, of course, complete nonsense. Like all intelligent and unbiased people I accept evolution as an immutable fact, although I have an idea that it sometimes works faster than is usually considered. If we take that as a given, then new animals are evolving all the time, and it makes sense that whenever a newly evolved animal is seen, but not catalogued by science, it should be grouped as a cryptid. However, it is the fallacious nature of the second of the principles outlined above that interests me most.

When I first came to England in 1971, and within a few months settled in North Devon, I was accompanied by a small library of books about British wildlife. Over the next six years until I left home in 1977, I spent much of my time watching, catching, keeping, and cataloguing the local wildlife. When I returned to North Devon nearly forty years later I found that the fauna of the area had in many ways changed beyond all recognition. From the books that I brought with me when I came to the UK for the first time I believed that the wildlife of my particular corner of our scepter'd isle would be like it said in the 1662 Book of Common Prayer; "As it was in the beginning, is now, and ever shall be, world without end. Amen". Well, I have yet to make my mind up about the kingdom of heaven, but this is certainly not the case in Woolsery.

The first difference is, obviously that the landscape, and therefore the mini biotopes that I used to explore as a child are largely gone. the little road between the village and Ashcroft Farm once had a dozen little ponds by the side of the road, each a Lilliputian world with its own unique ecosystem, and each with a surprisingly different population of water beetles, damselfly larvae and other invertebrates. Now they are all gone; victims of the diminishing water table and the increasing numbers of motor vehicles that traverse this once deserted little road every day.

There are many other examples, more than I care to mention because it is too depressing to complicate. But there are positive aspects to this change as well. In the six years I lived in the village during the 1970s, for example, I only saw one adult hawkmoth; *Sphinx ligustri,* which was found dying in the telephone box in the centre of the village by a local youth. On one occasion I found two caterpillars of *Deilephila elpenor* feeding on a fuchsia bush in the lane outside my house. Most excitingly, in 1974, a local farmer gave me a pupa he found buried in a field of seed potatoes. I tentatively identified it as *Acherontia atropos.*

The butterfly fauna has also changed remarkably over the past forty years, and I don't mean just in numbers. Various species that used to be common are far scarcer; the small copper *Lycaena phlaeas*, for example is very scarce here now, whereas once it was relatively common. But two other species which are now quite common, were considerably rarer back in the 1970s. I only ever saw brimstone (*Goneptryx rhamni*) once here during during the seventies (in the late summer of 1972), and I only saw orange tips (*Anthocharis cardamines*) rarely. Now both species are relatively common in the spring, and the former species can be seen in small numbers throughout the summer and even into the autumn.

Over the past ten years, partly as a result of this, I have become interested in the fluctuations of populations of different animal species.

In April 2011, I wrote this on the daily blog:

"Yesterday Graham and I went in search of an animal not seen in our part of North Devon for over thirty years. Like so many cryptozoological creatures, it is ethno-known, and our search was triggered by a chance encounter that a visiting fortean had with one of these animals.

Graham and I finally ran it down after an exciting car chase down twisting lanes, and we have conclusive proof of its existence.

I lived in Woolsery between 1971 and 1981, and most springtimes I would get on my bicycle and go up to the edges of Huddisford Woods (yes; the place where we got the big cat hairs last year), and look at the orange tips. *Anthocharis cardamines* is one of my favourite British butterflies, probably because it reminds me of my favourite butterfly from my childhood in Hong Kong. However, for some reason or another, they were never common in Woolsery, and began to die out in about 1977, and by the beginning of the decade that taste forgot, they had vanished."

When I described the species as being ethnoknown in the area, I was not being facetious. We had been told about them in passing by some visitors to the village, and I was very surprised because I had not seen them in the area for decades. But the casual visitors to the village were right and the expert lepidopterist completely wrong. There, I believe that you have a classic cryptozoological case in microcosm. And I think if we could understand more about the macropopulation dynamics of such a tiny creature, we could understand more about the nature of some more notorious cryptids.

Therefore, it seemed to be apposite for me to include a round up if just a few unusual British butterfly records that I believe to be of cryptozoological importance in this issue of our revamped journal. For the purposes of brevity I shall just discuss records from the last two butterfly years; 2013 and 2014.

Lampides boeticus, known in English as the long tiled blue is one of the commonest butterflies in the world, only not in the UK. It is found over an enormous range including Europe, Africa, South and Southeast Asia, and Australia. however, Central Europe is the northernmost limit of its range and it only very rarely reaches the UK. 2014 was a fairly typical *Lampides boeticus* year for us; there were only two records of which I am aware - one in Somerset in June and one in Hampshire a year later. However, 2013 was very different.

The excellent UK Butterflies website as this to say:

"The Long-tailed Blue (below) is an extremely rare migrant to the British Isles. It was first recorded from Brighton in East Sussex, and Christchurch in Hampshire, in August 1859. By 1939 a mere 36 sightings had been recorded - mostly of individuals. Between 1940 and 1988 another 85 sightings were recorded. The only major immigration was in 1945, a good year for migrants in general, when there were 38 sightings. A recent immigrant was observed in Hampshire in 2006.

However, the most noticeable influx occurred in 2013 when Long-tailed Blue were seen at 9 sites in Devon, Hampshire, Sussex, Kent and Suffolk. Mating pairs, eggs and larvae were also found, confirming that the species had successfully bred and, on 8th September 2013, the first of the offspring emerged in Wiltshire and Kent. Sightings from other counties followed, with sightings continuing into October."

It seems highly unlikely that this species will ever colonise the UK. Most of the migratory

species who regularly visit the United Kingdom either spend the winter hibernating as adults or larvae, or overwinter as eggs or pupae. However, *Lampides boeticus* is continually brooded, and does not hibernate through the winter months, which are - and look as if they will remain, unless climate change reaches truly catastrophic proportions - far too cold in the UK either to allow either any of the stages of this species to survive, or indeed its foodplant.

Sadly, therefore this beautiful little butterfly will remain a scarce visitor, unless helped over the winter months by human intervention, which would be compassionate but hardly ethical.

The changing weather patterns do, however, give us the possibility of some new arrivals on the British List. In his remarkable, and touching, book *The Butterfly Isles*, Patrick Barkham suggests that the Queen of Spain fritillary, is likely to be the next species to become a British resident, and gives a touching and poetic description of the courtship display of two of these lovely butterflies that he witnessed in southern England.

UK Butterflies has this to say about the species:

"This butterfly is an extremely rare immigrant to the British Isles with the first record from Gamlingay in Cambridgeshire in 1710. It was first noticed in numbers in 1818 and was seen every year until 1885 - with the highest total of 50 records in 1872. Since then, sightings are few and far between with an additional 42 records up until 1939. Between 1943 and 1950 an additional 75 records were added and, since then, there has again been a dearth of sightings with no sightings at all in some years. In 2009

several individuals were seen near the Sussex coast, including a sighting of a mating pair. Even so, there have been less than 400 sightings in total since it was first discovered.

Although females have been seen egg-laying, neither larvae nor pupae have been found in the wild except in the Channel Islands, where larvae were found in 1950, and larvae were again found in 1951 and 1957. However, in 1945, 25 individuals were recorded at Portreath in Cornwall, suggesting that a migrant female had deposited her eggs in the vicinity and that this concentration of adults were her offspring. Unfortunately, this species is unable to survive our winter. The vernacular name of "Queen of Spain" was given in 1775 by Moses Harris in *The Aurelian's Pocket Companion*, although no explanation for this name was given. This species is a rare migrant to the British Isles. The vast majority of sightings are from the south coast of England, with a fairly even spread from Cornwall to Kent. There are fewer records further north and several records from southern Ireland. It is believed that the presence of this species on our shores is dependent on individuals originating in northern France. Unfortunately, the number seen there is also decreasing due to loss of suitable habitat and this undoubtedly has a knock-on effect."

It has quite sensibly been hypothesised that over four hundred sightings of this rare insect could well be translated into about ten thousand individuals having been flying over the UK during 2013, which is an awesome concept, but quite a reasonable one when one examines the mathematics.

Whilst I would not disagree with Barkham's conclusions, I would personally think that there are two other species that are more likely to become established, both of which were once UK residents but have long been extirpated. There are also two mainland species which have never been British residents, which for two completely different reasons may become established in these islands. Each of these four species presents an interesting parallel for the cryptozoologist.

The Eurasian Swallowtail (*Papilio maechon*) pictured below, is another widely distributed butterfly. Although split into 37 currently recognised subspecies, and a number of races, it occurs throughout the Palearctic region in Europe and Asia; it also occurs across North America, and thus, is not restricted to the Old World, despite the common name.

In historical times two different species were found in the UK. The British subspecies *P.m.brittanicus* is a far more specialised insect than its continental cousin *P.m.gorganus*. Whereas the latter species eats a wide range of plants in the umbellifer family, the British subspecies only feeds upon one species of milk parsley, and is now confined to a very limited range within the Norfolk Broads. Gorganus, however, has turned up in Britain quite a lot in recent years, and has bred in both the last two years in various locations across southern Britain.

Whilst the advent of such a gorgeous new butterfly to the British list is undoubtedly something worth celebrating, it would, after all, be a returning native rather than an entirely new species for the country. There is a great deal of evidence suggesting that the gorganus subspecies was found quite widely in southern England until about two hundred years ago.

But if gorganus does become a British resident, whether or not it has been already in historical times, what are the implications for our own endemic subspecies. Is it doomed to eventual extinction by genetic dilution? If the European weather patterns are to carry on changing the way that they seem to be, it is quite a distinct possibility, at least to my mind, that gorganus will not only colonise southern England, but - in time - work its way up as far north as East Anglia, and if it does so, it would appear that contact between the two subspecies will be inevitable. Will they interbreed? And if they do, will it matter? I leave that moral conundrum to the conservationists of the latter half of this present century.

This has actually happened before, or at least it may well have done. In 1979 it was widely believed that one of the most peculiar British butterflies had become extinct. The large blue, *Maculinea arion* has a peculiarly elaborate lifecycle as described by UK Butterflies:

"The larva is parasitic in that it feeds on the grubs of a red ant, *Myrmica sabuleti*, on whom its existence depends. Although the dependence on ants had been known for many years, the dependence on a single species of ant, in order to maintain a viable population, was unknown to conservationists for many years until Jeremy Thomas discovered the association in the late 1970s. Unfortunately, the discovery came too late to save the native population. Today's reintroduction efforts focus as much on the population of ants present, as they do on the Large Blue itself."

A few years after the last official colony on Dartmoor was declared extinct, a complex and successful reintroduction took place. However, the insects introduced were from Sweden and of the nominate subspecies, and so, despite many people in North Devon, myself included, believing that there were still isolated colonies of the original British race of the large blue living at Blagdon Forest near Hartland, where I saw one in 1978 and at other isolated locations along the coast. Wherever the newly introduced butterflies reached any areas where the remnants of the original population were hanging on, then the original population would be destroyed by dilution.

In the case of *arion* it probably doesn't make any difference. The species was doomed in the UK, and if the people managing the new reintroduction programme jumped the gun a little, so what. Their meticulous and expensive programme of management at reserves such as Collard Hill in Somerset, was what made the introduction successful, but the fact does remain that if there were or are any isolated pockets of the original British insects still hanging on in isolated rural backwaters, then their DNA is doomed as soon as one of the more successful Swedish insects arrives on the scene.

The next of my interesting quasi-cryptozoological butterflies is the large tortoiseshell (*Nymphalis polychloros*) which has a range across Europe, northern Africa, and western Asia. It is quite similar in appearance to the better known small tortoiseshell (*Aglais urticae*) which is one of the best loved of British butterflies, but it is actually far closer related to the Camberwell Beauty (*Nymphalis antiopa*) an extremely rare visitor to our shores which the late L Hugh Newman suggested that often arrived in Britain accidentally in shipments of wood from Scandinavia.

The small tortoiseshell was very close to extinction in the UK a couple of years ago, although I am glad to say that it had made a fine recovery. Let us hope that it continues. This decline was initially blamed on the predations of a parasitic fly *Sturmia bella*, but when more work was carried out it appears that this was not the primary reason for the small tortoiseshell's decline.

However, something remarkably similar happened about a century ago when the population of *Nymphalis polychloros* which, until then, had been a reasonably well distributed and fairly common species across large swathes of the United Kingdom crashed, for reasons that remain obscure. If you examine the contemporary reports of people like P.B.W. Allan it appears that the species made something of a recovery in the 1930s but crashed again in the following decade, and was largely considered to be extirpated from the UK by the 1950s, even though it was not declared to be extinct until the 1980s.

There is a Fortean phenomenon that I have often noticed that no sooner has a major figure died than there are conspiracy theories emerging suggesting that they are not dead at all. The two most obvious examples of this are Elvis Presley and Adolf Hitler, both of whose departures from this world have been accompanied by such an enormous amount of conspiracy lore that it almost beggars belief. However, something that is less widely publicised is that the same thing happens with apparently extinct species. To once more choose the most obvious example of this, one has to look no further than the CFZ totem animal; the thylacine or Tasmanian wolf.

There have been so many eyewitness accounts of this animal since the last one supposedly died in 1936 that it would be a foolish man who claims that there is no hope of the species having survived.

However, exactly the opposite can also happen. I have always liked the book *Searching for Hidden Animals* by the late Roy Mackall. It is an intelligent and well written book, but it is full of all sorts of pieces of wild speculation, suggesting that even long extinct creatures like trilobites may have survived to the present day. This is what I have always thought of privately as 'The Science of Jumping to conclusions'. And both ends of the paradigm can, not particularly surprisingly, be found within the cryptozoological speculation about various species of British butterflies, past and present.

The best example of the more ludicrous end of this paradigm concern the occasional accounts of surviving specimens of the British race of the large copper (*Lycaena dispar*). UK Butterflies takes up the story:

"The Large Copper was first discovered from Dozen's Bank near Spalding in Lincolnshire in 1749. It became extinct in the British Isles in 1851 and was last recorded at Bottisham in Cambridgeshire. There is no doubt that the demise of this most spectacular butterfly was the result of changing fenland management and, in particular, the draining of the fens. On the continent this species lives in discrete colonies ranging from a few dozen adults to many hundred.

There have been several introduction attempts, the first at Woodwalton Fen, in Huntingdonshire, in 1927. On several occasions, the population had to be subsequently re-introduced or supplemented from captive stock. The British subspecies, dispar, was endemic to the British Isles and reintroductions have tended to use stock from the Netherlands, which is of the rare subspecies batavus. Unfortunately, all reintroduction attempts have ultimately failed. A project is being undertaken at Keele University to determine the feasibility of a Large Copper re-establishment programme in the British Isles. This species is extinct in the British Isles. Although the species was never widespread, it is believed that its former range also included Lincolnshire, Huntingdonshire, Norfolk and Somerset."

However, exactly the opposite is the case

when one examines the history of *Nymphalis polychloros* (see above) in the UK. There have been rare but consistent sightings pretty well every year since the creature was declared extinct. In the UK, for example the following records were made during 2014:

2014.08 16 Devon Hartland 1
2014.07.14 Essex Layer de la Haye 1 possible reported
2014.07.14 Suffolk Minsmere 1 possible reported
2014.07.13 Isle of Wight Bridlesthorpe Copse (private site) 1 reported
2014.03.30 Essex Basildon 1 reported
2014.03.14 Sussex Beckley Woods 1 reported
2014.03.13 Suffolk Peewit Lane, Felixstowe 1
2014.03.12 Suffolk Peewit Lane, Felixstowe 1
2014.03.10 Suffolk Peewit Lane, Felixstowe 1 reported

The Devonshire record, by the way was by our old friend Lars Thomas who saw a specimen on a garden wall in Hartland during the Weird Weekend over the third weekend of August. These records have been taken from the excellent Bugs Direct website run by Adrian Riley, and taken with thanks.

Pretty well every spring they are reported in the same stretch of woodland on the Isle of Wight, which would imply that they still exist here, and have chosen to overwinter in this particular wood. But there have been even more impressive reports of them having bred here.

However, this is where things begin to get a little strange. I am sure that I remember reading the source material for these claims a few years ago, but try as I can, I have been unable to find them anywhere. So I wrote the following letter on the forum of the excellent UK Butterflies website:

I remember reading within the past four or five years that a naturalist at either a National Trust or Natural England property in south Devon was claiming that large tortoiseshells had bred there. Does anyone have a link for this, or did I completely dream it?

Jon

And after some interesting discussions on the subject Dave Brown wrote back:

You were not dreaming, unless I shared your same dream. I recall reading about that, possibly the Branscombe area (Devon).

Being that no further news was put out the following year I assumed that no more were seen, or that it was deemed a release. Seems that the Isle of Wight is still the best chance.

In my reply I explained a little more about such internet anomalies:

Thank you Dave, it is nice to know that I haven't lost what little remains of my mind. This is one of the big problems with the Internet. Anyone can post anything, and whilst quite often things that are inaccurate do get removed, sometimes people do see them and a new 'meme' (if that is the correct word) enters the public consciousness.

A few years ago whilst researching for a book I was writing, for example, I came across what appeared to be a bona fide web page about the events just after the German Army surrendered in 1945, when - according to the author of this page - there was an unholy race between the British/US and the Soviet armies to get to the Danish border, and it was only because the British and Americans got there first that Denmark didn't join the Eastern Bloc. It was very plausible, but I have never found it again, and a Danish friend of mine who is a polymath had never heard of this.

So, Ladies and Gentlemen. Did the Large Tortoiseshell breed a few years ago at a National Trust property in Devon? Someone did write about it on a webpage that has since been taken down, because both Dave and I remember seeing it. But did the event actually happen? Or will it forever be one of those half truths that litter every branch of the natural sciences?

And that is where the matter rests as of the spring of 2015. There has only been one record for this year that has been logged with, what I believe to be the most trustworthy of the UK sightings websites, www.bugalert.net, and that was in Sussex earlier in April. Ironically, despite what I wrote above, 2015 has seen no reports from the Isle of Wight.

But there are is another tortoiseshell which we should now discuss. Tortoiseshells or angle winged butterflies are contained in the genera Nymphalis and Aglais, and are found across the northern hemisphere, and until very recently only two species had been recorded in Britain; the aforementioned small tortoiseshell and large tortoiseshell. A third species from eastern Europe, the yellow legged or scarce tortoiseshell *Aglais xanthomelas*, was known from one specimen from 1953. In recent years, however, it has expanded its range from eastern Europe into Scandinavia, and during 1994 there was another massive population explosion and butterflies of this species turned up in the low countries, and eventually in Britain.

Writing in *The Guardian* in mid March 2105, Patrick Barkham broke the news that many of us had been waiting for:

"One of the butterflies was seen again over two days last week in Holt country park, Norfolk, a sign it successfully hibernated over winter.

According to Butterfly Conservation's head of monitoring, Dr Tom Brereton, the woods near Holt provide suitable conditions for the butterfly, which feeds on sap from birch trees in the spring and lays eggs. If fine spring weather helps females and males to find each other to mate, the butterfly, also known as the yellow-legged tortoiseshell because of its distinctive straw-coloured legs, will breed in Britain for the first time.

"This new sighting is a truly historic event as it marks the first time this stunning butterfly has ever overwintered successfully in Britain," said Brereton. "We've been waiting apprehensively over the last couple of weeks for news to see if any scarce tortoiseshells would emerge from hibernation following last year's mini invasion. The butterfly prefers very cold winters and we weren't sure if any would survive our mild season. If more emerge as we head into spring, 2015 could see the first UK-born scarce tortoiseshells on record."

As far as that is concerned, we shall just have to wait and see.

The next butterfly has a long and chequered history in the UK. The map (*Araschnia levana*) is a creature of central and eastern Europe, but it has been expanding its range westward in recent years. The species is double, and occasionally triple, brooded and is unusual in that its two annual broods look very different (next column). The summer brood are black with white markings, looking like a miniature version of the white admiral and lacking most of the orange of the spring brood, which are superficially similar to commas. In the UK this species is considered to be a very rare vagrant, but there have also been several unsuccessful – and now illegal – attempts at introducing this species over the past 100 years or so: in the Wye Valley in 1912, the Wyre Forest in the 1920s, South Devon 1942, Worcester 1960s, Cheshire 1970s, South Midlands 1990s. All these introductions failed.

In the summer of 2014, I wrote that in an undisclosed location near Swanage in Dorset a number of European map butterflies (*Araschnia levana*) have been seen and photographed. This is not a native species to the UK, and is non migratory, but considering the fact that they are currently advancing their range in Europe, and the excellent summer we have had they could be the offspring of a vagrant female earlier in the year, rather than just having been released by a well meaning idiot. If they are natural vagrants has the UK got another new butterfly species in the offing? This species was deliberately introduced in 1912 when the butterfly became established in the Forest of Dean in Monmouthshire, and Symond's Yat in Herefordshire. The well-known entomologist A.B. Farn was so opposed to the deliberate introduction of a foreign species that, in 1914, he deliberately collected and destroyed every individual he could find. However, the ultimate demise of the colonies is believed to be the result of additional (and unknown) factors. One specimen was recorded in 1982 but apart from that, these appear to be the first British specimens.

But the provenance if these insects is uncertain. Despite the fact that the species is non-migratory, it is slowly but surely increasing its range, and it seems likely that it will arrive in the UK eventually. But are these specimens from Swanage the aforementioned bridgehead?

On 19th August Pete Eeles wrote on the UK Butterflies forum:

"Two days ago, an anonymous (to Joe Public) individual contacted BC to own up to an "accidental release". Unfortunately, the facts provided to BC don't stack up based on the observations made. So it could be a genuine release, or a miffed entomologist trying to muddy the waters because he or she didn't get to see the Map. So - not much of a conclusion really and, at this point in time, all we can say is that these Map are of "unknown origin" and, in all likelihood, a release (but it's a shame that the person releasing them didn't tell a convincing story - possibly because deliberately releasing them is breaking the law and they want to be given the benefit of the doubt)."

There has been speculation that the admission of a release is itself a hoax, attempting - with the best of possible intentions - to defuse interest in the nascent colony by the butterfly collecting community. yes there are still quite a few of them left, and specimens of one of the first ever European map butterflies to have been naturally in the UK would be worth a small fortune on eBay, whereas specimens that had been bred in the UK would be essentially worthless.

The first brood emerges in late April and May, but as - at the time of writing - it is already the end of April, and I have heard no reports of *Araschnia levana* this year, so, although we know that the Dorset butterflies produced eggs last summer, the suggestion that they all got either killed or captured by collectors seems to be a likely one. We can only live in hope.

These are only a few of the motile species within the world of British butterflies. There are other species such as the monarch, and the geranium bronze, which are native to geographical areas far removed from the western European habitats described in this article. Both have turned up in the UK regularly, but as neither milkweed (Asclepias spp) or pelargoniums (wrongly referred to as geraniums) are native to the UK, it would seem unlikely that they could become established, and until they do it would appear that the only way this occurrence could take place would be commensal to our own species, as both plants are only semi hardy and although kept widely in cultivation, will only survive the winter is in a greenhouse.

I have always maintained that the methodology first codified by Heuvelmans to define cryptozoology works just as well, if not better, for small species as it does for the larger and more spectacular creatures with which cryptozoology is more usually imagined to be dealing. My attitude has always been that cryptozoology is a little like Christianity: whether or not you believe the more supernatural aspects, or indeed the more extreme claims made by its devotees, it is still a pretty good set of rules by which to live your life. Or if you are a student of the natural sciences, a pretty good set of rules, by which to carry out one's investigations.

Keep watching the...ahem...skies.

DISCUSSION DOCUMENT: DUTIES FOR REGIONAL REPRESENTATIVES

The CFZ had had regional representatives for over twenty years now. Some of them have done remarkable things, some nothing at all, and some something in between. I originally intended my first wife to manage the list of regional reps, but as history shows, that never happened.

Ever since Alison and I split up I have been intending to ask someone else to take over the job, and finally a few months ago I got around to it. Ronan Coghlan has agreed to take over the onerous task, and has come up with a list of suggested roles

for regional representatives, which I post here for public discussion.

[1] In the event of a reported sighting of a mystery animal in the representative's area, all possible data should be gathered and forwarded to CFZ. Likewise, news of further developments should be sent on as they occur.

[2] Representatives should try to discover if there were any sightings or other anomalous events in their areas in the past, but should only send on stories of UFOs or ghosts if they consider them important, as otherwise their is the danger of CFZ being swamped.

[3] Representatives should, if possible, look into local folklore to discover if stories of anomalous events in the area occur. Liaisons should be initiated with the Bird, Butterfly and Conservation Officer in their areas where possible. They should, in addition, try to gather an archive of Fortean zoological material from their local studies libraries.

[4] Representatives should initiate liaisons with groups dealing with anomalies and nature in the area, provided they consider them and their personnel suitable.

[5] Representatives should have the option of offering sales of books to local bookshops. However, some might find this distasteful and so this should not be regarded as an actual representatives' duty.

A Colony of European Fire Salamanders in Northumberland?

Recently I was familiarising myself with some old haunts I used to frequent whilst growing up in the Northumbrian countryside and whilst visiting a pond in a quarry I used to 'play in' I was reminded of something that happened to me years ago.

The quarry is situated near Stocksfield in Northumberland and not far from my home village of West Mickley. The pond itself is man made and I remember it being crystal clear a week or so after filling as the water had settled. It is now completely naturalised and even has fish, water voles\rats, dragon flies etc. It is situated next to a much older marsh area which has water birds, insects etc. I seem to remember leeches too.

Anyway back to my memory. I think this would have been around the mid 1990s. One day I was mucking about near the pond and happened to lift a large flat log up to find what I 'think' were European salamanders. They had excavated some tunnels under the log and appeared to be a male and a female, as I don't suppose the species is sociable outside of mating. They seemed more textured than the *Neurergus strauchii* newt and had more rounded tails than the newt would have. On returning home I checked in a book, whilst my memory was fresh, and they did indeed seem to match the images of the salamander. They had a more 'string of inner tubes' look and did not have the finned tail that a newt would have. I did not touch the little creatures and so cannot say whether they had the throat colouring that the newt

Steve Baxter

would have had.

I returned with my camera shortly after but they either had been scared (accidentally by me) and moved on or were both out hunting etc.

I was young and could be mistaken, but if there is a colony of European salamanders living there I assume it would be an exciting find and may mean that the area comes under some kind of protection.

Upon remembering all this I did give the area a cursory examination this time but did not see evidence of the 'salamanders' but this being winter there was no chance.

I made my second visit to the area we discussed with the possibility of European salamander colony.

I didn't see anything this time, of course, it still being winter but did take some shots of the general area, the pond and the marsh. The pond had a small group of mallards which took flight as we approached. The area is frequented by the nearby farm's sheep who visit the pond and marsh to drink.

According to the farmer the land is in a bit of a bureaucratic grey area with no one actually taking responsibility for it. Seems strange.

He says that there are definitely fish in the pond, which species I have yet to find out and as the weather improves I will follow up with further visits.

I took a few images of the area so you can see the type of environment we are dealing with. I would expect the salamanders to be connected with the marsh rather than the pond. Again when weather improves I will visit again.

Letters

The editor and his compadres welcome letters for publication on all subjects covered by this magazine. However, we would like to stress that neither this magazine, or the CFZ are responsible for opinions expressed, which are purely those of the letter writer.

Gris Gris Gumbo Yaya

Hi Jon,

As a long-time Fortean, I share your frustration with the way that a subject which used to have aspirations to be a serious study of anomalous reports has become blurred beyond recognition by the modern tendency for hoaxes, popular entertainment and even jokes to be "shared" ad infinitum on Facebook and Twitter by casual internet users who can't be bothered to check the veracity of what they are sharing. The situation isn't helped by the fact that the mainstream media appear to judge the newsworthiness of Fortean-type stories not by their credibility but by their outrageousness.

A case in point was the "Whitstable giant crab" you mentioned on page 25 of Issue 52. This started out as a whimsical piece of artwork in issue 301 of *Fortean Times* (May 2013), which was a special tribute issue to the actor Peter Cushing on the hundredth anniversary of his birth. Whitstable was Cushing's adopted home, and the magazine included a lighthearted feature on "Weird Whitstable"

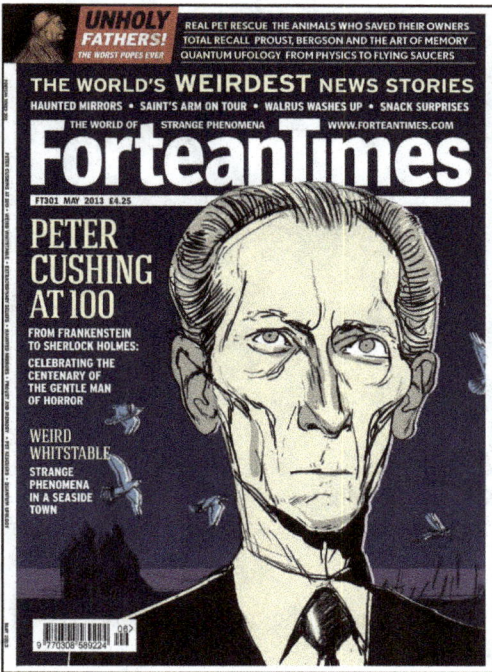

The article was the work of artist Quinton Winter, and the idea that a giant crab can be seen on Google Earth images of Whitstable was just one of his inventions. Sadly, however, it seems that people these days simply aren't wired up to understand irony. In October 2014, it was reported as a genuine news story in the *Sunday Express*, soon followed by several other tabloids

including the Daily Mail and the Daily Mirror. Within days, the story of "Crabzilla", as it became known, had gone viral on social media around the world!

Andrew May

(Source: *Fortean Times* issue 321, December 2014, page 2).

© Weird Whitstable

Walk on Gilded Splinters

Hi Jon,

I found this old print on a website which purports to show the fauna of the Americas.

I would have thought that some of this is from specimens, as the flying beast is surely a Jenny

Haniver?

http://bibliodyssey.blogspot.co.uk/2006/01/
americas-of-1671.html

Cheers

Steve Baxter

The Sun, the Moon and the Herbs

Jon,

I think I may have discovered something

quite curious. Not an undiscovered Captain Beefheart album, but something to suggest why a certain owlish character is where he is.

This site shows the underlying geology of the land:

**http://mapapps.bgs.ac.uk/
geologyofbritain/home.html**

If you look at the Mawnan area then something curious is seen.

Most of the area is fairly standard sedimentary rock, but there are a few smallish areas of intrusive dykes of igneous rock, including one directly under

Mawnan 'Bird-man' witnessed by June Melling 17 April (Easter Saturday) 76; (based on a sketch made by her the same day: DS/RJMR).

Mawnan Church, and one a couple of hundred yards south and east, under the nearby woods.

It seems that igneous rocks do something strange when stressed. Specifically, they become conductive in a very peculiar way reminiscent of how a semiconductor like doped silicon does.

http://www.scec.org/news/01news/ es_abstracts/Freund.pdf

There is a known association between earthquake lights and intrusive dykes in the general vicinity, which suggests that when the rocks of such an area are stressed, a lot of electricity is on the move and similar effects as the Hessdalen Lights are observed.

A quick Google search shows that the Falmouth area (conveniently close to Mawnan) does experience earthquakes, to a greater degree than the rest of that area, though a quick search doesn't seem to pull out the raw data for anywhere much.

Anyway, the general idea is that Mawnan Church is rather unusual geologically speaking, and the conditions to generate strange geophysical anomalies do exist at that place (and just that place).

Interesting, yes?

Dan H.

And a few days later he wrote:

I spotted something interesting. The main Owlman sightings took place in 1976, the year there was a drought.

Do you happen to have any record of the date and time of the known sightings that you could let me have, please? I would like to check them against historical tide data for that area (Falmouth ought to be close enough).

I'm thinking that the drought reduced the surface water to the extent that the physical loading on the rock strata on the landward area was reduced by several percent.

Then with the tides changing the loading on the seaward side, you have all the makings for flexing the rock strata in the area more than normal.

The key times for an Owlman sighting ought to be these:

- Prolonged drought plus high tide.
- Sudden, extremely heavy rain/snow and low tide.

These conditions must coincide with daylight between about 6:00AM and 20:00 PM, or there'll not be a suitable witness (i.e. juvenile female) around to see it.

Croker Courtbullion

Dear Readers,

Up until now the main mysteries about the Chinese Giant Salamander (given the name *Megalobatrachus sligoi* at the time, or Sligoi's Salamander) that turned up in a drain in 1922 in the grounds of the Hong Kong Zoological and Botanical Gardens, Mid-Levels, Hong Kong Island, and then taken to London Zoo, are not what was it, as it is now thought to be closely related to *Andreas davidianus*, but where exactly did it come from and why has it never been a part of Hong Kong's recognised fauna when it lives in Guangdong Province north of what used to be, prior to July 1997, the Colonial era border between British Hong Kong and the People's Republic of China? The last

Chinese Giant Salamander that I am aware of in this area was sighted in Shenzhen, (the metropolis that was a landscape of paddy fields in the 1920s and long afterwards) a few years ago.

To quote from a blog by Dr Darren Naish on the Edge of Existence site on December 8th 2010 concerning *Megalobatrachus sligoi*: "Edward G. Boulenger described it as representing the distinct species *Megalobatrachus sligoi*: it was supposedly distinct from the others on account of its smoother skin and flatter head (Boulenger 1924). It's been stated that additional specimens were later captured, though the details on these are extremely hazy. It's generally accepted today that Sligo's salamander is synonymous with *A. davidianus*, though I've been unable to discover whether the individual(s) concerned were possible natives or the result of introduction (let me know if you have more information); the species is missing from field guides to the Hong Kong herpetofauna (e.g., Karsen et al. 1986)."

On May 7th, after voting in Macclesfield I travelled to the Manchester University Main Library to use for the first time the *South China Morning Post* (S.C.M.P.) online newspaper database, which is not usually available to the public and cannot be accessed on the Net either. I found some information, in a letter to the S.C.M.P dated July 3rd 1953 from Alfred H. Crook, 79 Park Rd, Chiswick, London W.4., which I transcribe here:

Mr Crook reproduced two 25 Year Ago columns, the first one from July 11th 1908. The "Walking Fish" has aroused Mr A.H. Crook's sense of humour but in response to his request for more detail we have obtained one of the legs and will be happy to convince him thus far. The estimate of length was our own, but on measurement the fish was actually 2ft 4 inches. Captain Page informs us that the fish, reptile, missing link, etc grows to a length of several feet. There is no dorsal comb, but a semi-circular ridge. It is quite apparent, as Mr Crook states, that the fish is an amphibian, which may not be a contradiction in terms as Mr Crook suggests. If there was more interest in natural history in Hong Kong, steps would have been taken to secure specimens of the reptile for the museum, *or a live specimen for the Gardens.* (Emphasis my own. I am not categorically saying Mr Crook was responsible for secretly placing the giant salamander in the drain in the Gardens himself, but it can't be absolutely ruled out.)

The 1953 letter continues: twenty-five years ago. Extract from the S.C.M.P of July 14th 1908: To the Editor of the Morning Post. "so far only two facts have been gained about this animal – A: Its length 2ft 4in. B: That it has four limbs. We are left to surmise that it has a vertebral column. Yours etc Alfred H. Crook. The extracts arose from your remarks on what you called " A Walking Fish". The animal was ,of course, the giant salamander (*cryptobranchus maximus*). From enquiries as the result of our remarks I learned that there was not at that time a specimen at the Zoological Gardens in London.

(Mr Crook then describes his attempt to take two giant salamanders "from far up the West River", i.e. the Pearl River, and two large king crabs, *Limulus polyphemus*, from Tai O, Lantau, back to London in 1909, along with a Chinese pangolin and a "Japanese singing frog"". Called *Rona buergeri* in a Japanese book. "During vacations for over twenty years I tramped over various parts of the Japanese Islands from Kyushu in the south to Hokkaido in the north but I never came across this frog in its wild condition." But it is unclear whether Mr Crook ever did succeed in getting this mini zoo to London. However, and now I come to the crux of the matter, he writes in the same letter, still in 1909 presumably.

"I took the two salamanders and put them in a stream on the Peak which flowed from near Mountain Lodge and emptied itself into the sea near Pokfulam. Years afterwards I went to the Zoological Gardens here in London to see a giant salamander which was on exhibit there. The announcement said it had been obtained in Hong Kong. *I often wondered*

Jan Karlsson

Honey,
grab the
net !

*if it was one of the grand-children of the
ones I had let loose in the stream on the
Peak* (Emphasis my own.) I don`t think
there were any salamanders wild in Hong
Kong..."

Mountain Lodge was built in the 19th
Century and had a "second generation"
between 1900 and 1946. A camel once fell
to its death there, over a precipice. This
link here shows a photo of Mountain
Lodge in 1890: http://gwulo.com/
node/18194 .

I have a book called *Hong Kong Streets
and Places (Vol 1)* which shows streams
running down from Mountain Lodge into
what is now Pokfulam Reservoir. It is hard
to see how the salamanders could have
made it into another stream flowing in a

different direction, but maybe Mr Crook`s
memory is faulty? However, the reservoir
was built in 1877 and there was a conduit
running around to the north of Hong Kong
Island. So if this conduit stretched as far as
the Gardens, maybe the salamander found
its way thus? Conduit Rd is nearby after
all.

"Both *Megalobatrachus maximus* and *M.
sligoi* are now regarded as synonyms
of *Andrias davidianus*.

See   http://www.zsl.org/blogs/artefact-of-
the-month/celebrating-the-chinese-giant-
salamander.

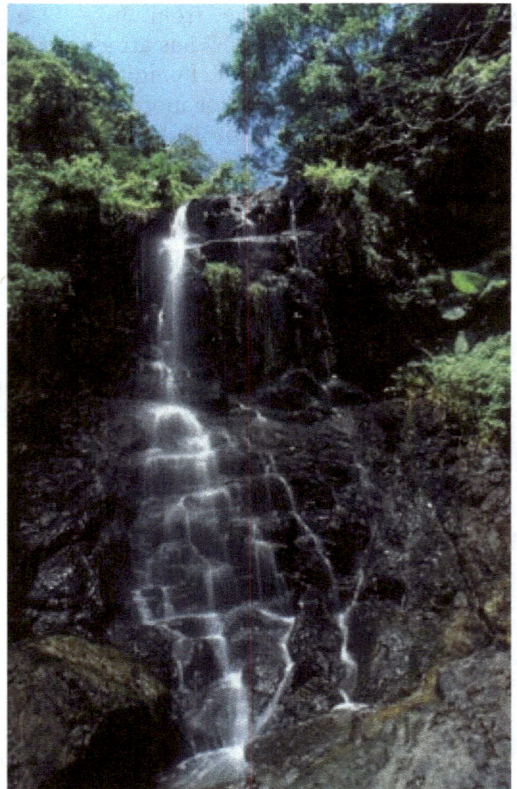

Other items I found included:

- A story of a primitive tadpole-like animal with a square head, no legs and a rat-like tail, found in a water pipe from Tai Tam. SCMP June 18th 1910.
- Crocodile in Kowloon Bay. SCMP July 14th 1910.
- Speculation that the Chinese porcupine was introduced from China. SCMP November 21st 1930.
- Two dholes seen near Lion Rock. SCMP March 18th 1936.
- Black animal, bear or porcupine, emerges from Byewash Reservoir. SCMP Jan 4th 1958.

REFERENCES :

'On a new giant salamander, living in the Society's Gardens' by E. G. Boulenger in 'Proceedings of the Zoological Society of London', 1924, pp. 173-174.
Karsen, S. J., Lau, M. W.-n. & Bogadek, A. 1986. *Hong Kong Amphibians and Reptiles*. Urban Council, Hong Kong.

Best Wishes

Richard Muirhead

Danse Fambeaux

Hey guys,

Hope you are both well. Just to give you a heads up that a lot has changed over in Sumatra. Some poor things are going on and we've just returned after someone sabotaged our expedition by putting the wrong words in the right ears within the authorities. Lots of embarrassed faces all round when they realised their mistake and they couldn't have been more apologetic but all our paperwork was in place, we'd spoken to the park officials on a couple of occasions and all was good with them.

Then someone had a chat and convinced them otherwise. Even without all of that going on there have been some massive changes with regards permits and permissions, such that we had to change our entire plan with only a couple of weeks before we left the UK.

Just don't want you guys to fall foul of the same situation with regards the permits etc - hence this message, I can't think that someone's trying to sabotage all expeditions as there was a guy up there just a couple of months ago at the lake with the full camera trap set up going on and he had no problems.

We could have just fallen foul of someone's ill intentions and greed rather than it being related to anything else.

Either way, the parks have increased massively the costs for using camera traps and the likes, possibly down to all the TV interest and the money to be made. Such that we had to leave all of our gear and change our plans.

The numbers we were hearing were that they want to charge people £1000 a month for a research permit. Suffice to say there's a lot of sh*t hitting the fan over there at the moment.

Anyway, stay safe, hope you're both well, hoping to make the Weird Weekend again this year, though that depends on work.

Peace and respect,

Sandy (Andrew Sanderson)

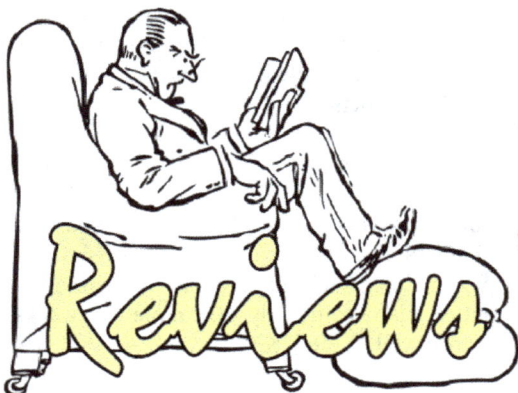

Reviews

MONSTERMIND

The magical life and art of
TONY 'DOC' SHIELS

RUPERT WHITE

Paperback: 157 pages
Publisher: Antenna Publications;
First edition (1 Feb. 2015)
Language: English
ISBN-10: 0993216412

It is very strange to be reviewing the biography of someone who you know pretty damn well. Tony Shiels and I have been friends since 1994, and there have been times during these years that I have spent a protracted length of time in his company. Despite what you may have heard, he is a dear, sweet man, and I truly am very fond of him. During the period of my divorce from my first wife, when I was hugely at odds with my parents, he was a father figure to me, and he still remains a figure of great importance in my life.

But outside my own personal experience of the man, he is a great painter, a peculiar playwright of some renown and, yes, he has been both a stage magician and the real thing. Do I think he can *do* magic of the thaumaturgical kind?

Yes I do, largely because I have seen him do it. So, we have established that he is a person worthy of having a biography written about him. The big question is, what is this book actually like? First of all, it is a slim volume with a heck of a lot of pictures. I am not quibbling about the pics, I was pleasantly surprised to find that there were a lot of pictures that I had not seen before, despite having a fairly large collection of Shielsiana of my own.

I am cited in the acknowledgements as having helped with the book, but it is a symptom of my rapidly hardening arteries that, although I remember swapping a few emails with the author, who seemed like a perfectly affable fellow, I cannot remember actually doing an interview or anything concrete to contribute. But no matter.

This is actually a pretty good book, and does pretty much what it says on the tin. It tells the bare bones story of Tony's remarkable life, and does so in a pleasingly concise manner. I learned several things about my old friend that

I didn't know before. This, as I have spent much of my time with him over the years in an earnest attempt to drink Cornwall dry, is quite possibly not surprising.

But, there is a big BUT. And the but is that this book singularly fails to capture the essence of this strange, often frightening, sometimes intimidating, occasionally frustrating, and always loveable man.

My first wife used to say that meeting Tony for the first time was like being hit by a hurricane, and in the past two and a bit decades I have not come across a description to better this one. I love the man very much, and have always thought that if I were going to write a book about him I should try to communicate that affection that I feel for him.

Indeed that was what I did in my 1997 book *The Owlman and Others* which dealt in great part with Tony and the events of 1976/7. I agree with the author of this present volume that "Doc" was merely a character invented by Tony for his own ends, and was not much more than a stage persona.

I have always considered it so, and I don't think I have either spoken to him or referred to him as 'Doc' any time this twenty years. But the author, like Mark Chorvinsky and everyone else who has tried to analyse this remarkable man in print, completely misses the point.

For Tony art, and magick, and stagecraft, and conjuring, and performance are all part of a surrealchemical whole in which he plays games with words, music, image and a whole slew of psychosocial smoke and mirrors to achieve....God knows what.

It is all part of The Case.

Greatly Surreal.

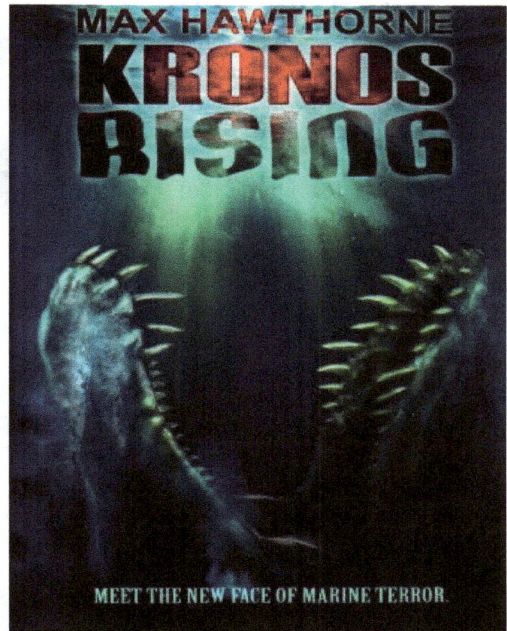

MEET THE NEW FACE OF MARINE TERROR.

Paperback: 562 pages
Publisher: Far From The Tree Press, LLC; 1 edition (2 April 2014)
Language: English
ISBN-10: 0615964958
ISBN-13: 978-0615964959

Hmmm. Back at Christmas 1975 someone (and I cannot for the life of me remember who) gave me a copy of *Jaws* by Peter Benchley, and the sixteen year old me enjoyed it massively, and it has been the benchmark by which I have judged all creaturefeature novels ever since. There is a whole subgenre of horror fiction dealing with 'real' creatures (albeit sometimes with fictional attributes) rending and slaying amongst the world of men, and although I would hardly describe myself as a fan of such books, I have read more than my fair share over the years. This book is actually quite a good example of the genre, although it has much the same plot as did Peter Benchley's book forty years ago. This time around, instead of great white sharks, it features a kronosaur. What? I hear you ask. Didn't they go extinct 100.5 million years ago? Well, um, yes.

Incidentally, for those of you of a cryptozoological bent, the whole idea that the prehistoric shark *Carcharadon megalodon* which was, of course a huge relative of today's great white shark, has survived to the present day instead of dying out 2.6mya was popularised by Benchley's book four decades ago. But I think that it is highly unlikely that this present volume will lead to a spate of scholarly articles suggesting that kronosaurs could have survived to the present day. The rationale given in *Kronos Rising* makes very little sense. According to the story, a pair of kronosaurs got trapped inside a volcanic crater by a tsunami caused by the meteorite impact that triggered the KT Extinction event. The fact that this occurred about 40 million years after the most recent known kronosaur fossil is conveniently ignored.

These kronosaurs, and a bevy of other prehistoric marine life lived for the next 65 million years in this enclosed crater until they were freed by seismic activity (8 think, I can't actually remember that bit) to do the aforementioned rending and slaying amongst the world of men. Ignoring the fact that the whole backstory of the book is somewhere between massively unlikely and nonsense, most of the book is actually a sensitively written and enjoyable read, even though it is as derivative as hell. The two main protagonists are very much 21st Century analogues of Benchley's characters, pressing all the right PC buttons. A sensitive recovering alcoholic smalltown sheriff and a staggeringly beautiful Japanese/Norwegian cetologist with a past as a green protestor as part of an organisation which sounds very much like Sea Shepherd. There are several satisfyingly nasty villains as well; a corrupt God bothering Senator, his oafish son, and a psychopathic big game hunter desperado who just happens to be the Japanese bint's estranged husband. It is all eminently entertaining.

The story rolls on quite predictably - the monster's predations have political and social ramifications far beyond what one might originally have expected, and eventually even the goody goody cetologist comes to terms with the fact that the nasty prehistoric survivors should be exterminated. I am not going to go through all of the plot twists because I don't want to spoil the book for any of you who decide to read it, and I do recommend it as an entertaining read. However, I now want to fast forward to the end of the book where, with two big cliches the author lets the side down massively. The whole book would, in fact, be considerably better if it were not for the final chapter. First of all, the author resorts to the literary device used by Dan Brown at the end of one of his books, when after having spent twenty odd chapters fighting the forces of evil the two protagonists then suddenly decide to have sex with each other, with no preamble, and with no real literary preparation except for the fact that the two protagonists are of opposite genders, and it was fairly obvious that with an eye for the possible film adaptation, Brown decided there needed to be a spot of rumpo in the book. Here the same thing happens, and totally out of the blue the two main protagonists start shagging like monkeys, for no particular reason.

And the final few pages are so predictable that I will not dignify them with a description. However, as long as you are prepared to suspend belief for a few hours, and ignore the last chapter and a bit, this is a surprisingly entertaining book, and far better written than most of its genre. JD

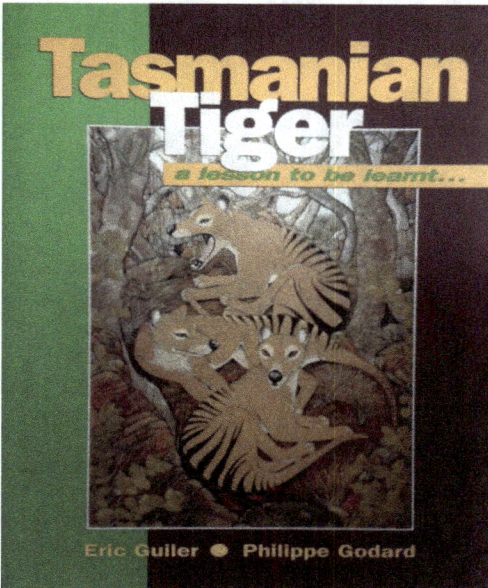

**Tasmanian Tiger: A Lesson To Be
Learnt
Eric Guiler and Philippe Godard
Abrolhos House 1998
ISBN 0958 579105**

Dr Eric Guiler of the University of Tasmania
spent decades looking for the thylacine but
unlike his colleague Col Bailey he was never
lucky enough to see one. He seemed slightly
bitter and this seeps through into his writings.
For all its sumptuous colour illustrations and
large hardback format Guiler and Godard's
book is not as good as Bailey's. That is not to
say it's a bad book, in fact it's very good but
it seems to lake the spark of Bailey's writing.
It's very much illustration led and there is an
interesting chapter on thylacine illustration
from the 1808 to the early 20th century. Many
of the early reconstructions were by artists
who had never seen a living animal and had
based their work on descriptions and poorly
preserved remains. The resulting paintings
resemble bear cubs, rats, possums and
placental wolves. As the decades passed the
artwork grew better as the creature became
more familiar. As with other works on the
subject there are chapters on the thylacine's
anatomy, habitat and the impact of white
colonization but there are other areas of study
rarely addressed elsewhere. A fascinating
chapter deals with thylacines in captivity both
in Australia / Tasmania and abroad. A
surprising number of these animals were kept
in zoos, London Zoo having the most.
However no attempt was ever made to start a
breeding programme. Thylacine were often
swapped for other animals and as they
became rarer commanded huge prices. As
late as the 1920s wild ones were being
captured for zoos.

Elsewhere the authors examine the
expeditions in search of this near legendary
species both official and unofficial. It is
interesting to read of the earl camera traps,
cumbersome and hugely unreliable. Lists of
thylacines captured at various sheep stations
and farms and the records of livestock losses.
Though thylacines did indeed prey on sheep
far more were lost to dogs, human rustlers
and poor management. The claims of some
politicians' were absurd, sighting numbers of
sheep kills that thylacines would have to have
been at plague proportions to have caused
them and it is clear they were not. Finally the
idea of cloning a thylacine is looked at in
detail and shown to be nothing more than a
tissue of nonsense. One prominent scientist
involved in this project once said that all the
modern thylacine sightings were reported
'when the pups closed' , showing he had not
even bothered to look at the data. It is
pleasing to see such an arrogant man exposed
as a charlatan. The future of the Thylacine is
in the wilds of Tasmania, New Guinea and
Australia not in a test tube. **Richard
Freeman**

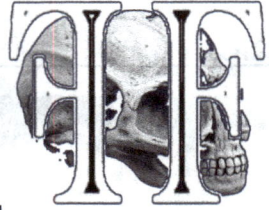

THE WORLD'S WEIRDEST PUBLISHING GROUP

We publish a lot of books. Indeed, I think that we could quite easily claim to be the world's foremost publishers of books about Fortean Zoology and allied disciplines, and our Fortean Words imprint is doing a great job in producing books on other non-zoological esoterica. However, I feel that it would be unethical to review our own titles. So here, to end this edition of *Animals & Men*, is a brief look at the books we have put out so far this year.

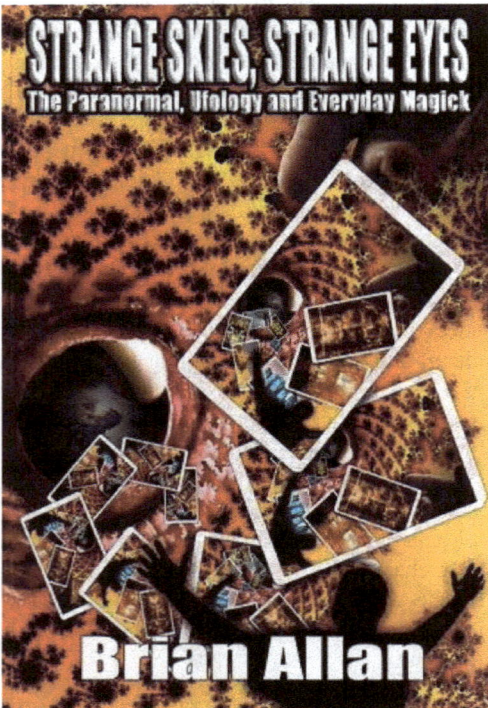

Paperback: 182 pages
Publisher: Fortean Words (21 Jan. 2015)
Language: English
ISBN-10: 1909488240
ISBN-13: 978-1909488243
Product Dimensions: 15.6 x 1 x 23.4 cm

This book could best be described as a series of reflections and observations, a distillation if you like, on the nature of, and relationship between, the occult as we encounter it in our daily lives. This includes the paranormal in all its guises and how this impinges upon the often fractious subject of Ufology, aspects of which may in themselves be other manifestations of the paranormal. From my own investigations and researches it seems to me that there is little if any difference between any of these subjects and where any divergences do occur, they are almost certainly in how different individuals experience and interpret the individual components.

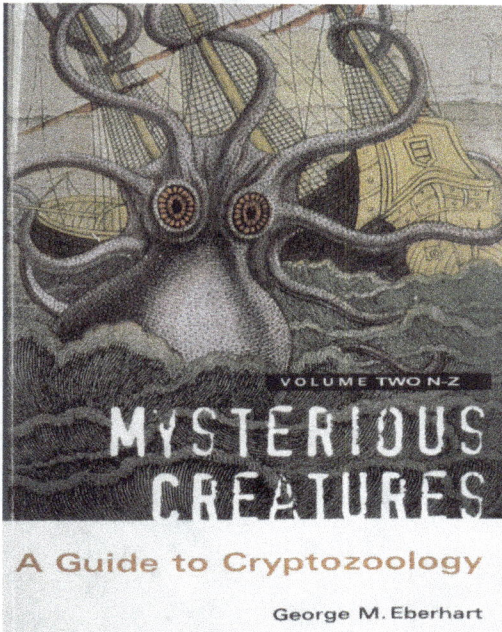

Paperback: 370 pages
Publisher: cfz (27 Jan. 2015)
Language: English
ISBN-10: 1909488259
ISBN-13: 978-1909488250
Product Dimensions: 18.9 x 2 x 24.6 cm

In the hi-tech, industrialised world of the 21st Century, monsters still loom large. Far from being relegated to the realm of myth and legend, mysterious creatures seem to be alive and well today.

In this, the second of two volumes, acclaimed researcher George M. Eberhart provides a comprehensive list of the creatures that roam our monster-haunted planet.

Covering 'N-Z', Volume Two tackles some well-known beasts as well as more esoteric creatures few will have heard of.

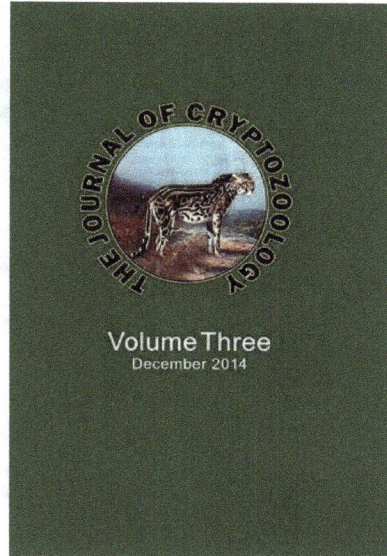

Paperback: 100 pages
Publisher: cfz (8 April 2015)
Language: English
ISBN-10: 1909488275
ISBN-13: 978-1909488274
Product Dimensions: 15.6 x 0.5 x 23.4 cm

Following the demise of Cryptozoology (published by the now-defunct International Society of Cryptozoology), there has been no peer-reviewed scientific journal devoted to cryptozoology for quite some time. Consequently, the Journal of Cryptozoology has been launched to remedy this situation and fill a notable gap in the literature of cryptids and their investigation. For although some mainstream zoological journals are beginning to show slightly less reluctance than before to publish papers with a cryptozoological theme, it is still by no means an easy task for such papers to gain acceptance, and, as a result, potentially significant, serious contributions to the subject are not receiving the scientific attention that they deserve. Now, however, they have a journal of their own once again.

weird weekend **2015**

The Small School, Hartland, North Devon
www.cfz.org.uk

August 14 - 16 2015
TEL: +44 (0) 1237 431413

Three days of Monsters and Mysteries

For the second year running......

HARTLAND, YOU'VE NEVER HAD IT SO WEIRD